Lecture Notes in Artificial Intelligence   6301

Edited by R. Goebel, J. Siekmann, and W. Wahlster

Subseries of Lecture Notes in Computer Science

W0193328

Thomas Sturm   Christoph Zengler (Eds.)

Automated Deduction in Geometry

7th International Workshop, ADG 2008
Shanghai, China, September 22-24, 2008
Revised Papers

 Springer

Series Editors

Randy Goebel, University of Alberta, Edmonton, Canada
Jörg Siekmann, University of Saarland, Saarbrücken, Germany
Wolfgang Wahlster, DFKI and University of Saarland, Saarbrücken, Germany

Volume Editors

Thomas Sturm
Max-Planck-Institut für Informatik, RG 1: Automation of Logic
66123 Saarbrücken, Germany
E-mail: sturm@mpi-inf.mpg.de

Christoph Zengler
Universität Tübingen
Wilhelm-Schickard-Institut für Informatik
Symbolic Computation Group
72076 Tübingen, Germany
E-mail: christoph@zengler.eu

ISSN 0302-9743　　　　　　　　　　e-ISSN 1611-3349
ISBN 978-3-642-21045-7　　　　　　 e-ISBN 978-3-642-21046-4
DOI 10.1007/978-3-642-21046-4
Springer Heidelberg Dordrecht London New York

Library of Congress Control Number: 2011926785

CR Subject Classification (1998): I.2.3, I.3.5, F.4.1, G.2-3, D.2.4

LNCS Sublibrary: SL 7 – Artificial Intelligence

Typesetting: Camera-ready by author, data conversion by Scientific Publishing Services, Chennai, India

Printed on acid-free paper

Springer is part of Springer Science+Business Media (www.springer.com)

Preface

Automated Deduction in Geometry (ADG) 2008 continued an established and fruitful series of biannual international workshops in that area. Previous meetings have taken place in Toulouse (1996), Beijing (1998), Zurich (2000), Linz (2002), Gainesville (2004), and Pontevedra (2006). The seventh workshop was held at the East China Normal University (ECNU) in Shanghai during September 22–24, 2008. ADG 2008 was co-organized by the CAS-MPG Partner Institute for Computational Biology, the Shanghai Institute of Biology Sciences, and the Chinese Academy of Sciences.

While the ADG workshops themselves are quite open also for the informal discussion of work in progress, the selected contributions for the proceedings generally undergo a very thorough and highly selective reviewing process. This publication in the LNAI series of selected papers from 2008 continues a tradition established with the first ADG in 1996.

As the Chair of the Program Committee, I would like to thank in the first place all PC members, who are listed on the next page, for their competence and dedication during two refereeing processes: first for the workshop contributions and then for the selected papers published here; all this comprised a considerable timespan. Zhenbing Zeng did a perfect job with the local organization at the exceptionally interesting location in Shanghai. Manual Kauers and Christoph Zengler greatly supported me with the organization of the online reviewing process and the preparation of these proceedings. Jesús Escribano and Miguel Abánades created and maintained a beautiful and informative website, which was most helpful for the organizers as well as for the attendees of the workshop. Last but not least, I particularly want to thank Tomás Recio and Dongming Wang for their advice that has accompanied my organizational work for more than two years.

March 2011   Thomas Sturm

Organization

Invited Speakers

Shang-Ching Chou   Wichita State University, USA/Zhejiang
                                      University, China

Tetsuo Ida   University of Tsukuba, Japan

Organizing Committee

Zhengbing Zeng   Shanghai, China, Chair
Miguel A. Abánades   Madrid, Spain
Jesús Escribano   Madrid, Spain
Christoph Zengler   Tübingen, Germany

Program Committee

Thomas Sturm   MPI-INF, Saarbrücken, Germany, Chair
Hirokazu Anai   Fujitsu Lab, Japan
Francisco Botana   Pontevedra, Spain
Christopher Brown   Annapolis, USA
Giorgio Dalzotto   Pisa, Italy
Jacques Fleuriot   Edinburgh, UK
Xiao-Shan Gao   Beijing, China
Hoon Hong   Raleigh, USA
Deepak Kapur   Albuquerque, USA
Manuel Kauers   Linz, Austria
Montserrat Manubens   Barcelona, Spain
Pavel Pech   Ceske Budejovice, Czech Republic
Tomás Recio   Santander, Spain
Georg Regensburger   Linz, Austria
Jürgen Richter-Gebert   Munich, Germany
Pascal Schreck   Strasbourg, France
Meera Sitharam   Gainesville, USA)
Dongming Wang   Beijing, China/Paris, France
Min Wu   Shanghai, China
Bican Xia   Beijing, China
Zhenbing Zeng   Shanghai, China

Table of Contents

Contributed Papers

Dynamical Systems of Simplices in Dimension Two or Three

Gérald Bourgeois[1] and Sébastien Orange[2]

[1] GAATI, Université de la polynésie française, BP 6570, 98702 FAA'A, Tahiti,
Polynésie Française
bourgeois.gerald@gmail.com

[2] LMAH, Université du Havre, 25 rue Philippe Lebon, 76600 Le Havre, France
Sebastien.Orange@univ-lehavre.fr

Abstract. Let $T_0 = (A_0 B_0 C_0 D_0)$ be a tetrahedron, G_0 be its centroid and S be its circumsphere. Let (A_1, B_1, C_1, D_1) be the points where S intersects the lines $(G_0 A_0, G_0 B_0, G_0 C_0, G_0 D_0)$ and T_1 be the tetrahedron $(A_1 B_1 C_1 D_1)$. By iterating this construction, a discrete dynamical system of tetrahedra (T_i) is built. The even and odd subsequences of (T_i) converge to two isosceles tetrahedra with at least a geometric speed. Moreover, we give an explicit expression of the lengths of the edges of the limit. We study the similar problem where T_0 is a planar cyclic quadrilateral. Then (T_i) converges to a rectangle with at least geometric speed. Finally, we consider the case where T_0 is a triangle. Then the even and odd subsequences of (T_i) converge to two equilateral triangles with at least a quadratic speed. The proofs are largely algebraic and use Gröbner bases computations.

Keywords: Dynamical systems, Gröbner basis, Tetrahedron.
MSC: Primary 51F, 13P10, 37B.

1 Introduction

Notation. Let M, N be two points of $\mathbb{R}^d$. The Euclidean norm of the vector $\overrightarrow{MN}$ is denoted by MN.

1.1 The General Problem

Let $T_0 = (A_{0,0} \cdots A_{0,d})$ be a d-simplex, G_0 be its centroid, S be its circumsphere in $\mathbb{R}^d$, O be the center of S. For all $0 \leq i \leq d$ the line $G_0 A_{0,i}$ intersects S in two points $A_{0,i}$ and $A_{1,i}$. Let T_1 be the d-simplex $(A_{1,0} \cdots A_{1,d})$, and G_1 be its centroid. If we iterate this process then we produce a dynamical system of d-simplices $(T_i)_{i \in \mathbb{N}}$ with centroids $(G_i)_{i \in \mathbb{N}}$.

Using Maple, numerical investigations with thousands of random simplices in dimensions up to 20 indicate that the following results seem true[1]:

---

[1] Private communication with A. Edmonds (Indiana University).

T. Sturm and C. Zengler (Eds.): ADG 2008, LNAI 6301, pp. 1–21, 2011.
© Springer-Verlag Berlin Heidelberg 2011

1. The sequence $(OG_i)_{i\in\mathbb{N}}$ is decreasing and tends to 0.
2. The sequences $(\mathcal{T}_{2i})_{i\in\mathbb{N}}$ and $(\mathcal{T}_{2i+1})_{i\in\mathbb{N}}$ converge to two d-simplices with centroid O.

The questions *1.* and *2.* are known, but it seems to us that no literature is available about these problems in the case $d \geq 3$. The case $d = 2$ was considered in [2] by the first author. Here we give an other proof, that is essentially algebraic. Moreover, we obtain a better estimate for OG_n. In this article, we prove also, and this is the main result, the assertions *1.* and *2.* in the case $d = 3$. Moreover, similar results are proved for cyclic quadrilaterals as a degenerate case of the previous one.

1.2   The Case $d = 3$

The tetrahedron $\mathcal{T}_0$ is not planar. Therefore, for every $i \in \mathbb{N}$, $\mathcal{T}_i = (A_i B_i C_i D_i)$ is a tetrahedron such that its four vertices are not coplanar. These vertices are ordered and $\mathcal{T}_i$ is viewed as a point of $\mathcal{S}^4$. Let ϕ (resp. ψ) be the function which transforms $\mathcal{T}_0$ into $\mathcal{T}_1$ (resp. G_1).

Remark 1.

1. $\mathcal{T}_1$ is the image of $\mathcal{T}_0$ by an inversion that leaves $\mathcal{S}$ globally invariant.
2. ϕ admits no non-planar tetrahedron as fixed point.
3. If $\mathcal{T}$ is an isosceles tetrahedron then $\mathcal{T}$ is a fixed point of $\phi \circ \phi$.

Our main results are as follows.

The sequences $(\mathcal{T}_{2i})_{i\in\mathbb{N}}$ and $(\mathcal{T}_{2i+1})_{i\in\mathbb{N}}$ are well-defined and converge, with at least geometric speed. Their limits are two non-planar isosceles tetrahedra that are symmetric with respect to O. We consider also the degenerate case where $\mathcal{T}_0$ is a planar cyclic convex quadrilateral. Then, the sequence $(\mathcal{T}_i)_{i\in\mathbb{N}}$ converges to a rectangle, with at least geometric speed. If $\mathcal{T}_0$ is a harmonic quadrilateral then the limit is a square. We conjecture that the convergence is with order three. Moreover, in both cases, we give explicit expressions for the lengths of the edges of the limit from the ones of $\mathcal{T}_0$.

1.3   The Case $d = 2$

The triangle $\mathcal{T}_0$ is not flat. Therefore for every $i \in \mathbb{N}$, $\mathcal{T}_i = (A_i B_i C_i)$ is a triangle such that its vertices are pairwise distinct.

We prove that the sequences $(\mathcal{T}_{2i})_{i\in\mathbb{N}}$ and $(\mathcal{T}_{2i+1})_{i\in\mathbb{N}}$ are well-defined and converge to two equilateral triangles that are symmetric with respect to O. Moreover, these sequences converge with at least quadratic speed.

Numerical investigations show that the convergence is much faster in dimension two than in dimension three.

1.4   Computational Aspects of the Proofs

Along this paper, we use rational functions of the square of the lengths of the edges of the d-simplices. Naturally some systems of polynomial equations appear. Computations with such systems require Gröbner bases softwares. They are performed by using the J. C. Faugere's software "FGb"[2] and the computer algebra system MAPLE.

Considering a fixed geometrical configuration, several authors used Gröbner bases for solving geometric constraints (see [6] and [9]). On the other hand, we do not know any paper using Gröbner bases in order to study discrete dynamical systems in Euclidean geometry.

2 Standard Definitions and Results about Tetrahedra

2.1 General Tetrahedra

Let $\mathcal{T} = (ABCD)$ be a tetrahedron and G be its centroid and $\mathcal{V}(\mathcal{T})$ be its volume.

Notations, assumptions:

1. In the case where A, B, C, D are not coplanar, we denote by O the center of the circumsphere of $\mathcal{S}$. Moreover we assume that its radius is 1.
2. In the case where A, B, C, D are coplanar, we assume that $ABCD$ is a cyclic quadrilateral. We denote by O the center of the circumcircle of $\mathcal{T}$. Moreover, we assume that its radius is 1.
3. Let $BC = a$, $CA = b$, $AB = c$, $AD = a'$, $BD = b'$, $CD = c'$. Note that the 3-tuple (a, b, c) does not play the same role as (a', b', c').

Proposition 2. *We have*

1. $OG^2 = 1 - \dfrac{1}{16}(a^2 + b^2 + c^2 + a'^2 + b'^2 + c'^2)$

   $ = 1 - \dfrac{1}{4}(GA^2 + GB^2 + GC^2 + GD^2).$

2. $GA^2 = \dfrac{3}{16}(a'^2 + b^2 + c^2) - \dfrac{1}{16}(a^2 + b'^2 + c'^2),$

   $GB^2 = \dfrac{3}{16}(b'^2 + c^2 + a^2) - \dfrac{1}{16}(b^2 + c'^2 + a'^2),$

   $GC^2 = \dfrac{3}{16}(c'^2 + a^2 + b^2) - \dfrac{1}{16}(c^2 + a'^2 + b'^2),$

   $GD^2 = \dfrac{3}{16}(a'^2 + b'^2 + c'^2) - \dfrac{1}{16}(a^2 + b^2 + c^2).$

Proof. 1. See [1, p. 64].
   2. See [7].   □

Notations.

1. For all $u, v, w \in \mathbb{R}^3$, $Gram(u, v, w)$ denotes the Gram matrix of these three vectors.

---

[2] J.C. Faugère, LIP6, Paris 6. The FGb software can be downloaded from the website: www-calfor.lip6.fr/∼ jcf/. It is included in Maple 13.

2. We denote by $\Gamma(A, B, C, D)$ the Cayley-Menger determinant

$$\det \begin{pmatrix} 0 & 1 & 1 & 1 & 1 \\ 1 & 0 & c^2 & b^2 & a'^2 \\ 1 & c^2 & 0 & a^2 & b'^2 \\ 1 & b^2 & a^2 & 0 & c'^2 \\ 1 & a'^2 & b'^2 & c'^2 & 0 \end{pmatrix}.$$

3. We denote by $\Delta(A, B, C, D)$ the determinant $\det \begin{pmatrix} 0 & c^2 & b^2 & a'^2 \\ c^2 & 0 & a^2 & b'^2 \\ b^2 & a^2 & 0 & c'^2 \\ a'^2 & b'^2 & c'^2 & 0 \end{pmatrix}.$

Proposition 3. *We have:*

1. $288 \times \mathcal{V}(\mathcal{T})^2 = 8 \times \det \left(Gram \left(\overrightarrow{AB}, \overrightarrow{AC}, \overrightarrow{AD} \right) \right) = \Gamma(A, B, C, D)$. *Therefore,*
   A, B, C, D *are not coplanar if and only if* $\Gamma(A, B, C, D) > 0$.
2. $\Delta(A, B, C, D) = -2\Gamma(A, B, C, D)$.
3. $\det \begin{pmatrix} 1/2 & 1 & 1 & 1 & 1 \\ 1 & 0 & c^2 & b^2 & a'^2 \\ 1 & c^2 & 0 & a^2 & b'^2 \\ 1 & b^2 & a^2 & 0 & c'^2 \\ 1 & a'^2 & b'^2 & c'^2 & 0 \end{pmatrix} = 0.$

Proof. *1.* and *2.* See [7].
   3. Follows directly from *2.*   □

Proposition 4. *The expression* $Pt(\mathcal{T}) = -\Delta(A, B, C, D)$ *has the following properties:*

1. $Pt(\mathcal{T}) = (bb' + cc' - aa')(cc' + aa' - bb')(aa' + bb' - cc')(aa' + bb' + cc')$
   $\qquad = 2a^2a'^2b^2b'^2 + 2b^2b'^2c^2c'^2 + 2c^2c'^2a^2a'^2 - a^4a'^4 - b^4b'^4 - c^4c'^4.$
2. *If* $\mathcal{T}$ *is not planar then each factor of the product appearing in 1. is positive*
   (Ptolemy's inequality). Moreover $Pt(\mathcal{T}) = 576 \times \mathcal{V}(\mathcal{T})^2$.
3. *If* $\mathcal{T}$ *is a cyclic quadrilateral then* $Pt(\mathcal{T}) = 0$.

Proof. See [3] or [7].   □

Remark 5. Recall the relation:

$$16 \times S^2 = 2a^2b^2 + 2b^2c^2 + 2c^2a^2 - a^4 - b^4 - c^4, \tag{1}$$

where S is the area of a triangle with lengths of the edges a, b, c.

2.2 Isosceles Tetrahedra

Definition 6. *The tetrahedron* $\mathcal{T}$ *is said to be* isosceles (or equifacetal) *if*

$$a = a', \ b = b' \ and \ c = c'.$$

Remark 7. A planar isosceles tetrahedron is a rectangle.

Proposition 8.   *1. The tetrahedron T is isosceles if and only if $G = O$.*
   2. If T is isosceles then $72\, \mathcal{V}^2(T) = (b^2 + c^2 - a^2)(c^2 + a^2 - b^2)(a^2 + b^2 - c^2)$
      and $a^2 \leq b^2 + c^2$, $b^2 \leq c^2 + a^2$, $c^2 \leq a^2 + b^2$.
   3. One has $aa' + bb' + cc' \leq 8$. Moreover, T is isosceles if and only if

$$aa' + bb' + cc' = 8\,.$$

Proof. See [7].   □

The following result is a direct consequence of Proposition 8 and Proposition 2.

Proposition 9. *The tetrahedron T is isosceles if and only if*

$$GA = GB = GC = GD\,.$$

3   Deformation from T_0 to T_1 ($d = 3$)

3.1   Parameters and Notations

We adapt our previous notations. Let $n \in \mathbb{N}$ and A_n, B_n, C_n, D_n be the vertices of the tetrahedron T_n. We assume only that A_n, B_n, C_n, D_n are not all equal. Then the functions ϕ, ψ are continuous in T_n.

Definition 10. *The quantities $d_{12}^n = A_n B_n{}^2$, $d_{13}^n = A_n C_n{}^2$, $d_{14}^n = A_n D_n{}^2$, $d_{23}^n = B_n C_n{}^2$, $d_{24}^n = B_n D_n{}^2$, $d_{34}^n = C_n D_n{}^2$ are called the parameters of the tetrahedron T_n.*

Notations.

1. To simplify the notations, in the case $n = 0$, we denote the parameters $d_{ij}^0{}'s$
   by the $d_{ij}'s$ ($1 \leq i < j \leq 4$).
2. We put $g_1 = G_0 A_0{}^2$, $g_2 = G_0 B_0{}^2$, $g_3 = G_0 C_0{}^2$ and $g_4 = G_0 D_0{}^2$.
3. The opposites of the powers of the points G_0 and G_1 with respect to $\mathcal{S}$ are
   denoted by

$$p_0 = 1 - OG_0{}^2 = \frac{1}{4}(g_1 + g_2 + g_3 + g_4) \text{ and } p_1\,.$$

According to Proposition 2, the $g_i's$ ($1 \leq i \leq 4$) and p_0 are polynomial functions of the $d_{ij}'s$ ($1 \leq i < j \leq 4$).

Proposition 11.   *1. For all $1 \leq i < j \leq 4$, $d_{ij}^1 = p_0{}^2 \dfrac{d_{ij}}{g_i g_j}$.*

2. The equality $p_1 = \dfrac{p_0{}^2}{16} \displaystyle\sum_{1 \leq i < j \leq 4} \dfrac{d_{ij}}{g_i g_j}$ holds.

3. $OG_0{}^2$ and $OG_1{}^2$ are rational functions of the $d_{ij}'s$ ($1 \leq i < j \leq 4$).

Proof. Assertion *1.* Clearly, $G_0 A_0 \times G_0 A_1 = p_0$ and $G_0 A_1{}^2 = \dfrac{p_0{}^2}{g_1}$. Since the

triangles $(G_0 A_0 B_0)$ and $(G_0 B_1 A_1)$ are similar, one has $\dfrac{A_1 B_1}{A_0 B_0} = \dfrac{G_0 A_1}{G_0 B_0}$ and

$A_1 B_1{}^2 = d_{12} \dfrac{p_0{}^2}{g_1 g_2}$.
   Finally, assertion *2.* follows from *1.* and *3.* follows from *2.*   □

3.2 Inequalities

The first key point is the following.

Theorem 12. *We have the inequalities:*

1. $OG_1 \leq OG_0$,
2. *for all* $(ijkl) \in \{(1234), (1324), (1423)\}$, $d_{ij}d_{kl} \leq d_{ij}^1 d_{kl}^1$,
3. $Pt(\mathcal{T}_0) \leq Pt(\mathcal{T}_1)$.

Moreover, if one of these inequalities is an equality then $\mathcal{T}_0$ is an isosceles tetrahedron.

Proof. 1. One may assume that $g_1 \geq g_2 \geq g_3 \geq g_4 > 0$. We replace the variables g_1, g_2, g_3, g_4 with s_1, s_2, s_3, s_4 that are defined as follows:

$$s_1 = g_1 - g_2, \; s_2 = g_2 - g_3, \; s_3 = g_3 - g_4 \text{ and } s_4 = \frac{1}{g_4}.$$

Obviously, $s_1, s_2, s_3 \geq 0$ and $s_4 > 0$. Put $E = 64 \dfrac{(p_1 - p_0)g_1 g_2 g_3}{p_0}$.

The signum of E is equal to the signum of $OG_0{}^2 - OG_1{}^2$.

Using the J. C. Faugere's software "FGb", and the polynomial expressions of the g_i's $(1 \leq i \leq 4)$ in the d_{ij}'s $(1 \leq i < j \leq 4)$, we compute the Gröbner basis of the system:

$$[s_1 - g_1 + g_2, \; s_2 - g_2 + g_3, \; s_3 - g_3 + g_4, \; s_4 g_4 - 1]$$

for the lexicographical order induced by

$$d_{12} > d_{13} > d_{14} > d_{23} > d_{24} > d_{34} > s_1 > s_2 > s_3 > s_4.$$

The normal form of E with respect to this Gröbner basis returns this quite surprising result:

$$
\begin{aligned}
E = & \, d_{23}(s_1 - s_3)^2 + 16 s_2^2 s_3^2 s_4 + 4 d_{34} s_1 s_2 + 12 s_1 s_2^2 + 4 s_1 s_2 s_3 + 20 s_2 s_3^2 + 21 s_2 s_3^3 s_4 + \\
& \, d_{24} s_3^2 + d_{34} s_2^2 + 4 s_1 s_3^2 + d_{24} s_1^2 s_3 s_4 + d_{34} s_1^2 s_3 s_4 + 12 s_3^3 + 2 d_{24} s_1 s_2 s_3 s_4 + 8 s_2^3 + \\
& \, 3 d_{24} s_1 s_3^2 s_4 + 6 d_{34} s_1 s_2 s_3 s_4 + 4 d_{34} s_2^2 + d_{34} s_3^2 + 5 d_{34} s_2^2 s_3 s_4 + d_{24} s_1^2 + 3 d_{34} s_1 s_2^2 s_4 + \\
& \, 4 s_2^3 s_3 s_4 + 3 d_{34} s_2 s_3^2 s_4 + 2 d_{24} s_1 s_3 + 12 s_2^2 s_3 + 2 d_{34} s_1 s_3 + 4 d_{34} s_2 s_3 + 9 s_3^4 s_4 + 2 d_{34} s_3^3 s_4 + \\
& \, d_{34} s_1^2 s_2 s_4 + s_1^2 s_2 s_3 s_4 + 4 s_1 s_2^2 s_3 s_4 + s_1^2 s_3^2 s_4 + 10 s_1 s_2 s_3^2 s_4 + 6 s_1 s_3^3 s_4 + 3 d_{34} s_1 s_3^2 s_4.
\end{aligned}
$$

Obviously, E is a non-negative real. Moreover if $E = 0$ then $s_1 = s_2 = s_3 = 0$ and, according to Proposition 9, $\mathcal{T}_0$ is an isosceles tetrahedron.

2. Let $\Lambda = \dfrac{p_0^4}{g_1 g_2 g_3 g_4}$. By Proposition 11, $d_{ij}^1 d_{kl}^1 = \Lambda d_{ij} d_{kl}$. According to the inequality of arithmetic and geometric means (AGM), one has $\Lambda \geq 1$ and $d_{ij}d_{kl} \leq d_{ij}^1 d_{kl}^1$.

Moreover, if $d_{ij}d_{kl} = d_{ij}^1 d_{kl}^1$ then $\Lambda = 1$. Again, by AGM inequality, we have $g_1 = g_2 = g_3 = g_4$ and $\mathcal{T}_0$ is an isosceles tetrahedron.

3. A computation gives $Pt(\mathcal{T}_1) = \Lambda^2 Pt(\mathcal{T}_0)$ and we reason as in 2. $\square$

4 Solution of the Case $d = 3$. Part 1

In this section we prove that the cluster points of the sequence $(\mathcal{T}_i)_{i \in \mathbb{N}}$ are isometric.

Theorem 13. *1. The sequence $(OG_n)_{n \in \mathbb{N}}$ converges to 0.*

2. Let $\mathcal{T} = (ABCD)$ be a cluster point of the bounded sequence $(\mathcal{T}_i)_{i \in \mathbb{N}}$. Then $\mathcal{T}$ is not planar and is isometric to a fixed isosceles tetrahedron with the following parameters:
$d_{12}^{\infty 2} = L d_{12} d_{34}, d_{13}^{\infty 2} = L d_{13} d_{24}, d_{14}^{\infty 2} = L d_{14} d_{23}$, where

$$L = \frac{64}{(\sqrt{d_{12} d_{34}} + \sqrt{d_{13} d_{24}} + \sqrt{d_{14} d_{23}})^2}.$$

Proof. According to Theorem 12.*3.*, the sequence $(Pt(\mathcal{T}_i))_{i \in \mathbb{N}}$ is increasing and positive. For all $i \in \mathbb{N}$, the four points A_i, B_i, C_i, D_i are not coplanar. Since for all $i \in \mathbb{N}$, $G_i \notin \{A_i, B_i, C_i, D_i\}$, the sequence $(\mathcal{T}_i)_{i \in \mathbb{N}}$ is well-defined and the functions ϕ, ψ are continuous in the $\mathcal{T}_i's$ ($i \in \mathbb{N}$). Moreover, the bounded sequence $(Pt(\mathcal{T}_i))_{i \in \mathbb{N}}$ converges to a real number $Pt_0 > 0$.

By Theorem 12.*1.*, the bounded sequence $(OG_n)_{n \in \mathbb{N}}$ is decreasing and converges to $r \geq 0$. We can extract a subsequence $(\mathcal{T}_{n_k})_{k \in \mathbb{N}}$ from $(\mathcal{T}_n)_{n \in \mathbb{N}}$ such that the sequences $(A_{n_k})_{k \in \mathbb{N}}$, $(B_{n_k})_{k \in \mathbb{N}}$, $(C_{n_k})_{k \in \mathbb{N}}$, $(D_{n_k})_{k \in \mathbb{N}}$ converge to A, B, C, D. Let $\mathcal{T}$ be the tetrahedron $(ABCD)$ and G be its centroid. Therefore $(G_{n_k})_{k \in \mathbb{N}}$ converges to G, and we have $OG = r$ and $Pt(\mathcal{T}) = Pt_0$. Hence, A, B, C, D are not coplanar and $\phi(\mathcal{T}) = \mathcal{T}' = (A'B'C'D')$ is well-defined. Let G' be its centroid.

Assume that $\mathcal{T}$ is not isosceles. Then, Theorem 12.*1.* implies that $OG' < OG$. Let $\epsilon \in (0, OG - OG')$. Consider $\alpha > 0$ such that $OG_{n+1} - OG' < \epsilon$ as soon as $||\mathcal{T} - \mathcal{T}_n|| < \alpha$. There exists $n_k \in \mathbb{N}$ such that $||\mathcal{T} - \mathcal{T}_{n_k}|| < \alpha$. One has $OG_{n_k+1} - OG' < \epsilon$ and consequently $OG_{n_k+1} < OG = r$. That is a contradiction. Therefore, $\mathcal{T}$ is isosceles and $(OG_n)_{n \in \mathbb{N}}$ converges to 0. For every $1 \leq i < j \leq 4$, the parameter d_{ij}^∞ of $\mathcal{T}$ is the limit of the sequence $(d_{ij}^{n_k})_{k \in \mathbb{N}}$.

According to the proof of Theorem 12.*1.*, for a permutation $\{i, j, k, l\}$ of $\{1, 2, 3, 4\}$, the bounded sequence $(d_{ij}^n d_{kl}^n)_{n \in \mathbb{N}}$ satisfies the equality

$$d_{ij}^{n+1} d_{kl}^{n+1} = \Lambda_n d_{ij}^n d_{kl}^n.$$

The sequence $(d_{ij}^n d_{kl}^n)_{n \in \mathbb{N}}$ is increasing and convergent. Thus, the infinite product $\Pi_{n=0}^\infty \Lambda_n$ converges to $L \geq 1$, satisfying the equality $d_{ij}^{\infty 2} = L d_{ij} d_{kl}$. The relation $d_{12}^\infty + d_{13}^\infty + d_{14}^\infty = 8$ gives the explicit value of L as a function in the $d_{ij}'s$ ($1 \leq i < j \leq 4$). Thus all cluster points have same parameters. □

Corollary 14. *For every $1 \leq i < j \leq 4$, the sequence $(d_{ij}^n)_{n \in \mathbb{N}}$ converges to d_{ij}^∞.*

Proof. Let $\mathcal{U} = \{(1234), (1324), (1423)\}$. For $(ijkl) \in \mathcal{U}$, the sequence $(d_{ij}^n d_{kl}^n)_{n \in \mathbb{N}}$ converges to $d_{ij}^{\infty 2}$ and the sequence $\left(\sum_{(ijkl) \in \mathcal{U}} \sqrt{d_{ij}^n d_{kl}^n}\right)_{n \in \mathbb{N}}$ converges to 8. From the inequality

$$2 \sum_{(ijkl) \in \mathcal{U}} \sqrt{d_{ij}^n d_{kl}^n} \leq \sum_{i<j} d_{ij}^n \leq 16,$$

we deduce that the sequence $\left(\sum_{(ijkl)\in\mathcal{U}}\left(\sqrt{d_{ij}^n}-\sqrt{d_{kl}^n}\right)^2\right)_{n\in\mathbb{N}}$ converges to 0 and, for $(i,j,k,l)\in\mathcal{U}$, the sequence $(d_{ij}^n-d_{kl}^n)_{n\in\mathbb{N}}$ converges to 0 too. Therefore the sequences $(d_{ij}^n)_{n\in\mathbb{N}}$ and $(d_{kl}^n)_{n\in\mathbb{N}}$ converge to d_{ij}^∞. $\square$

It remains to prove that the sequence $(\mathcal{T}_{2i})_{i\in\mathbb{N}}$ cannot rotate around O.

5   Solution of the Case $d = 3$. Part 2

We assume that $\mathcal{T}_0$ is not isosceles. We study the convergence speed of the sequence $(OG_n{}^2)_{n\in\mathbb{N}}$.

Definition 15. *Let $(f_n)_{n\in\mathbb{N}}, (g_n)_{n\in\mathbb{N}}$ be two positive sequences. We say that $f_n = \Theta(g_n)$ if and only if there exist two positive reals α, β such that, for all n large enough, $\alpha g_n \leq f_n \leq \beta g_n$.*

5.1   Taylor Series I

For all $n \in \mathbb{N}$ and $1 \leq i < j \leq 4$, let $h_{ij}^n = d_{ij}^\infty - d_{ij}^n$, $h_n = (h_{ij}^n)_{1\leq i<j\leq 4}$ and $\delta_n = \sum_{1\leq i<j\leq 4} h_{ij}^n$. For all $n \in \mathbb{N}$, one puts

$$\epsilon_n = (h_{12}^n - h_{34}^n)^2 d_{12}^\infty + (h_{13}^n - h_{24}^n)^2 d_{13}^\infty + (h_{14}^n - h_{23}^n)^2 d_{14}^\infty.$$

Remark 16. The sequence $(h_n)_{n\in\mathbb{N}}$ converges to zero.

Proposition 17. *The equality $OG_{n+1}{}^2 = \dfrac{-\delta_n}{16} - \dfrac{1}{16^2}\epsilon_n + O(\|h_n\|^3)$ holds.*

Proof. Recall that $OG_n{}^2 = \dfrac{-\delta_n}{16}$. A computation done with the software "FGb" gives the following expansion:

$$OG_{n+1}{}^2 = \frac{-(\delta_n + \tau_n)}{16\left(1 + \dfrac{\delta_n}{4}\right)} + O\left(\|h_n\|^3\right), \tag{2}$$

$$= \frac{-\delta_n}{16} - \frac{1}{16}\left(\tau_n - \frac{\delta_n{}^2}{4}\right) + O\left(\|h_n\|^3\right), \tag{3}$$

where

$$\tau_n - \frac{\delta_n{}^2}{4} = \frac{(h_{12}^n - h_{34}^n)^2 d_{12}^\infty + (h_{13}^n - h_{24}^n)^2 d_{13}^\infty + (h_{14}^n - h_{23}^n)^2 d_{14}^\infty}{16} + O\left(\|h_n\|^3\right).$$

$\square$

5.2   Taylor Series II

Recall that $d_{14}^\infty = 8 - d_{12}^\infty - d_{13}^\infty$ and $d_{12}^\infty + d_{13}^\infty > 4$. The second key point is the following:

Proposition 18. *1. One has $\delta_n = O(||h_n||^2)$.*
2. There exists $k > 0$ such that $\epsilon_n \geq -k\delta_n$, for all n large enough.

Proof. 1. From Proposition 3.2., the tetrahedron $\mathcal{T}_n$ satisfies the algebraic equality

$$Eq : \quad \Delta(A_n, B_n, C,, D_n) + 2\Gamma(A_n, B_n, C_n, D_n) = 0.$$

With Maple, we obtain the Taylor expansion in h_n of the LHS of (Eq) at order 2:

$$4(4 - d_{12}^\infty)(4 - d_{13}^\infty)(d_{12}^\infty + d_{13}^\infty - 4)\delta_n + \sigma'_n + O(||h_n||^3) \text{ where } \sigma'_n = O(||h_n||^2) \text{ is a}$$

non-negative quadratic form in the h_{ij}'s, $(1 \leq i < j \leq 4)$. Hence $\delta_n = O(||h_n||^2)$.
2. Let $h'_n = (h_{13}^n, h_{14}^n, h_{23}^n, h_{24}^n, h_{34}^n) \in \mathbb{R}^5$ and

$$\sigma_n(h'_n) = \sigma'_n(-h_{13}^n - h_{14}^n - h_{23}^n - h_{24}^n - h_{34}^n, h'_n).$$

One has the equality:

$$\sigma_n = -4(4 - d_{12}^\infty)(4 - d_{13}^\infty)(d_{12}^\infty + d_{13}^\infty - 4)\delta_n + O(||h_n||^3).$$

Moreover, σ_n is a quadratic form in h'_n, whose symmetric associated matrix Σ is defined by:

$$\Sigma = \begin{pmatrix} \Sigma_{11} & \Sigma_{12} & \Sigma_{12} & \Sigma_{14} & 2\Sigma_{12} \\ * & \Sigma_{22} & 2\Sigma_{12} - \Sigma_{22} & \Sigma_{12} & 2\Sigma_{12} \\ * & * & \Sigma_{22} & \Sigma_{12} & 2\Sigma_{12} \\ * & * & * & \Sigma_{11} & 2\Sigma_{12} \\ * & * & * & * & 4\Sigma_{12} \end{pmatrix}$$

with
$$\begin{cases} \Sigma_{11} = 2d_{12}^\infty d_{13}^\infty + 32 - 12d_{13}^\infty + d_{13}^{\infty 2} - 12d_{12}^\infty + d_{12}^{\infty 2}, \\ \Sigma_{12} = -4d_{12}^\infty + 16 + d_{12}^\infty d_{13}^\infty - 8d_{13}^\infty + d_{13}^{\infty 2}, \\ \Sigma_{14} = -d_{12}^{\infty 2} + 4d_{12}^\infty + d_{13}^{\infty 2} - 4d_{13}^\infty, \\ \Sigma_{22} = -4d_{13}^\infty + d_{13}^{\infty 2}. \end{cases}$$

In the same way, we put:

$$\epsilon'_n(h'_n) = \epsilon_n(-h_{13}^n - h_{14}^n - h_{23}^n - h_{24}^n - h_{34}^n, h'_n).$$

One has $\epsilon_n = \epsilon'_n + O(||h_n||^3)$. As previously, ϵ'_n is a quadratic form in h'_n, whose symmetric associated matrix is:

$$E = \begin{pmatrix} E_{11} & E_{12} & E_{12} & E_{14} & 2E_{12} \\ * & E_{22} & 2E_{12} - E_{22} & E_{12} & 2E_{12} \\ * & * & E_{22} & E_{12} & 2E_{12} \\ * & * & * & E_{11} & 2E_{12} \\ * & * & * & * & 4E_{12} \end{pmatrix} \text{ with } \begin{cases} E_{11} = d_{12}^\infty + d_{13}^\infty, \\ E_{12} = d_{12}^\infty, \\ E_{14} = d_{12}^\infty - d_{13}^\infty, \\ E_{22} = 8 - d_{13}^\infty. \end{cases}$$

Let F be the vector space:

$$F = \left\{ [x_1, x_2, x_3, x_4, x_5]^T \in \mathbb{R}^5 \mid x_1 = x_4, x_2 = x_3, x_5 = -x_1 - x_2 \right\}.$$

Clearly, F is the kernel of the matrices Σ and E.

Definition 19. *We say that a vector $h_n \in \mathbb{R}^6$ has an "acceptable" value if the quantities $\left(\sqrt{d_{ij}^\infty + h_{ij}^n}\right)_{1 \leq i < j \leq 4}$ satisfy the relation in Proposition 3.3.*

Assume that h_n has an acceptable value and a small norm. An easy computation proves that $h_n' \in F$ if and only if $\mathcal{T}_n$ is an isosceles tetrahedron. We know that if $\mathcal{T}_0$ is not isosceles then, for every $n \in \mathbb{N}$, the tetrahedron $\mathcal{T}_n$ is not isosceles. Therefore, $h_n' \notin F$. Let $F^\perp$ be the orthogonal of F for the usual scalar product. Write $h_n' = u_n + v_n$, where $u_n \in F$ and $v_n \in F^\perp$. The sequence $(v_n)_{n \in \mathbb{N}}$ converges to 0 and, for all $n \in \mathbb{N}$, $v_n \neq 0$. Moreover, one has $h_n'^T \Sigma h_n' = v_n^T \Sigma v_n > 0$ and $h_n'^T E h_n' = v_n^T E v_n > 0$.

A computation gives:

$$Spec(E_{|F^\perp}) = \{2d_{13}^\infty, 8d_{12}^\infty, 2d_{14}^\infty\}$$

and $Spec(\Sigma_{|F^\perp}) = \{2(4 - d_{12}^\infty)(4 - d_{14}^\infty), 8(4 - d_{13}^\infty)(4 - d_{14}^\infty), 2(4 - d_{12}^\infty)(4 - d_{13}^\infty)\}$,

with associated eigenvectors:

$$[-1, 0, 0, 1, 0]^T, [1, 1, 1, 1, 2]^T, [0, -1, 1, 0, 0]^T.$$

Remark 20. The matrices E and Σ are simultaneously diagonalizable.

Let $m = \min\left\{ \dfrac{d_{13}^\infty}{(4 - d_{12}^\infty)(4 - d_{14}^\infty)}, \dfrac{d_{12}^\infty}{(4 - d_{13}^\infty)(4 - d_{14}^\infty)}, \dfrac{d_{14}^\infty}{(4 - d_{12}^\infty)(4 - d_{13}^\infty)} \right\}$,

$M = \max\left\{ \dfrac{d_{13}^\infty}{(4 - d_{12}^\infty)(4 - d_{14}^\infty)}, \dfrac{d_{12}^\infty}{(4 - d_{13}^\infty)(4 - d_{14}^\infty)}, \dfrac{d_{14}^\infty}{(4 - d_{12}^\infty)(4 - d_{13}^\infty)} \right\}$ and

$$\rho = \min\left\{ d_{13}^\infty(4 - d_{13}^\infty), d_{12}^\infty(4 - d_{12}^\infty), d_{14}^\infty(4 - d_{14}^\infty) \right\}.$$

If h_n has an acceptable value, then $m \leq \dfrac{\epsilon_n'(h_n')}{\sigma_n(h_n')} \leq M$ and

$$\epsilon_n' = \epsilon_n + O\left(||h_n||^3\right) \geq m\sigma_n = -4m(4 - d_{12}^\infty)(4 - d_{13}^\infty)(4 - d_{14}^\infty)\delta_n + O\left(||h_n||^3\right).$$

To conclude, it remains to show the following property:

$$\epsilon_n = \Theta\left(||h_n||^2\right) \text{ and } -\delta_n = \Theta\left(||h_n||^2\right) \quad (*).$$

Indeed, the property $(*)$ implies that for all large enough n, $\epsilon_n > -k\delta_n$, where $k \in \mathbb{R}$ such that $0 \leq k < 4\rho$. We postpone the proof of the property $(*)$ in the next section.   □

Remark 21.   1. If the limit is a regular tetrahedron then $m = M = \dfrac{3}{2}$.

2. It can be proved that $\rho \leq \dfrac{32}{9}$, with equality if and only if the limit is a regular tetrahedron.

5.3 Proof of the Property (∗)

Lemma 22. *If $(ijkl) \in \mathcal{U} = \{(1234), (1324), (1423)\}$, then*

$$d^n_{ij} d^n_{kl} - (d^\infty_{ij})^2 = O\left(||h_n||^2\right).$$

Proof. Let $L_n = \dfrac{64}{(\sqrt{d^n_{12}d^n_{34}} + \sqrt{d^n_{13}d^n_{24}} + \sqrt{d^n_{14}d^n_{23}})^2}$. Obviously, L_n converges to 1.

We have $(d^\infty_{ij})^2 = L_n d^n_{ij} d^n_{kl}$ and $d^n_{ij} d^n_{kl} - (d^\infty_{ij})^2 = \left(\dfrac{1}{L_n} - 1\right)(d^\infty_{ij})^2$, and consequently,

$$\frac{1}{L_n} - 1 = \frac{(\sqrt{d^n_{12}d^n_{34}} + \sqrt{d^n_{13}d^n_{24}} + \sqrt{d^n_{14}d^n_{23}})^2 - 8^2}{64}.$$

Thus, if we put $u_n = \sqrt{d^n_{12}d^n_{34}} + \sqrt{d^n_{13}d^n_{24}} + \sqrt{d^n_{14}d^n_{23}} - 8$, then we have to prove that $u_n = O\left(||h_n||^2\right)$. One has

$$u_n = \sum_{(ijkl) \in \mathcal{U}} \sqrt{(d^\infty_{ij})^2 + d^\infty_{ij}(h^n_{ij} + h^n_{kl}) + O\left(||h_n||^2\right)} - 8,$$

$$= \sum_{(ijkl) \in \mathcal{U}} d^\infty_{ij} \sqrt{1 + \frac{h^n_{ij} + h^n_{kl}}{d^\infty_{ij}} + O\left(||h_n||^2\right)} - 8,$$

$$= \sum_{(ijkl) \in \mathcal{U}} d^\infty_{ij} + \frac{h^n_{ij} + h^n_{kl}}{2} + O(||h_n||^2) - 8 = \frac{\delta_n}{2} + O\left(||h_n||^2\right),$$

$$= O\left(||h_n||^2\right),$$

by Proposition 18. □

Lemma 23. *If $(ijkl) \in \mathcal{U}$, then $h^n_{ij} + h^n_{kl} = O(||h_n||^2)$.*

Proof. One has

$$d^n_{ij} d^n_{kl} - (d^\infty_{ij})^2 = d^\infty_{ij}(h^n_{ij} + h^n_{kl}) + h^n_{ij} h^n_{kl} = O\left(||h_n||^2\right)$$

by Lemma 22. □

Lemma 24. *For n large enough, there exists $(ijkl) \in \mathcal{U}$ such that*

$$|h^n_{ij} - h^n_{kl}| \geq \frac{1}{\sqrt{6}} ||h_n||.$$

Proof. By Lemma 23, there exists $A > 0$ such that, for all $n \in \mathbb{N}$ and $(ijkl) \in \mathcal{U}$,

$$|h^n_{ij} + h^n_{kl}| \leq A ||h_n||^2.$$

Let $\epsilon \in \left(0, \dfrac{1}{6A}\right)$. There exists $N \in \mathbb{N}$ such that, for all $n \geq N$, one has $||h_n|| < \epsilon$. Let $n \geq N$ be an integer. We may assume that $|h^n_{12}| = \sup_{i<j} |h^n_{ij}|$. Thus, $|h^n_{12}| \geq \dfrac{||h_n||}{\sqrt{6}} \geq \dfrac{|h^n_{12}|}{\sqrt{6}}$ and $|h^n_{12} + h^n_{34}| \leq 6A(h^n_{12})^2$ and

$$|h_{12}^n - h_{34}^n| \geq 2|h_{12}^n| - |h_{12}^n + h_{34}^n| \geq 2|h_{12}^n| - 6A(h_{12}^n)^2.$$

Since $0 < |h_{12}^n| \leq ||h_n|| < \epsilon < \dfrac{1}{6A}$, we deduce that $6A(h_{12}^n)^2 < |h_{12}^n|$ and $|h_{12}^n - h_{34}^n| \geq |h_{12}^n| \geq \dfrac{1}{\sqrt{6}}||h_n||$. $\qquad \square$

Proposition 25. *If $(h_n)_{n\in\mathbb{N}}$ has acceptable values, then for all n large enough, $\epsilon_n \geq \lambda||h_n||^2$ with $\lambda = \dfrac{\inf_{i<j} d_{ij}^\infty}{6}$. Moreover, one has the equality:*

$$-\delta_n = \Theta\left(||h_n||^2\right).$$

Proof. The equality $\epsilon_n = (h_{12}^n - h_{34}^n)^2 d_{12}^\infty + (h_{13}^n - h_{24}^n)^2 d_{13}^\infty + (h_{14}^n - h_{23}^n)^2 d_{14}^\infty$ and Lemma 24 give the first part. If h_n has an acceptable value then, by Proposition 18,

$$\epsilon_n' = \epsilon_n + O\left(||h_n||^3\right).$$

Hence, by the first part, if $\mu < \lambda$ then for all n large enough, $\epsilon_n' \geq \mu||h_n||^2$. For all $n \in \mathbb{N}$, one has $\epsilon_n'(h_n') \leq M\sigma_n(h_n')$. As a consequence, for all n large enough, $\sigma_n \geq \dfrac{\mu}{M}||h_n||^2$.

Recall that (see proof of Proposition 18):

$$-\delta_n = \frac{\sigma_n}{\nu} + O\left(||h_n||^3\right) \quad \text{with} \quad \nu = 4(4 - d_{12}^\infty)(4 - d_{13}^\infty)(4 - d_{14}^\infty).$$

Consequently if $\mu_1 < \lambda$, then for all n large enough, $-\delta_n \geq \dfrac{\mu_1}{M\nu}||h_n||^2$. $\qquad \square$

5.4 The Main Result in Dimension Three

Let $r = \max\left\{\dfrac{|d_{12}^\infty - 2|}{2}, \dfrac{|d_{13}^\infty - 2|}{2}, \dfrac{|d_{14}^\infty - 2|}{2}\right\} \in \left[\dfrac{1}{3}, 1\right)$.

Theorem 26. *The sequences $(\mathcal{T}_{2i})_{i\in\mathbb{N}}$ and $(\mathcal{T}_{2i+1})_{i\in\mathbb{N}}$ are well-defined and converge to two non-planar isosceles tetrahedra that are symmetric with respect to O. Moreover, the convergence is with at least geometric speed.*

Proof. By the Proposition 25, one has:

$$OG_{n+1}{}^2 = \frac{-\delta_n}{16} - \frac{1}{16^2}\epsilon_n + O\left(||h_n||^3\right)$$

$$< \frac{-\delta_n}{16} + \frac{1}{16^2}k\delta_n + O\left(||h_n||^3\right) \sim OG_n{}^2\left(1 - \frac{k}{16}\right).$$

Finally, for all n large enough, $OG_{n+1} \leq q\, OG_n$, where $q > \sqrt{1 - \dfrac{\rho}{4}} = r$.

Let $\mathcal{B}_n$ be the closed ball of center O and radius OG_n. Let α_n be the positive angle $((A_{n+1}A_n), (A_{n+1}A_{n+2}))$. Let l_n be the distance between the two sets $\mathcal{B}_n \cap (A_{n+1}A_n)$ and $\mathcal{B}_n \cap (A_{n+1}A_{n+2})$. Obviously, the sequence $(\alpha_n)_{n\in\mathbb{N}}$ converges to 0 and $G_n G_{n+1} \geq l_n$. We see easily that $l_n \sim \alpha_n$ and $A_n A_{n+2} \sim 2\alpha_n$. Thus, for all n large enough, $A_n A_{n+2} \leq 3G_n G_{n+1}$.

Let $p \in \mathbb{N}^*$. For all n large enough, one has:

$$A_n A_{n+2p} \leq 3 \sum_{k=n}^{n+2p-2} G_k G_{k+1} \leq 3 \sum_{k=n}^{n+2p-2} (OG_k + OG_{k+1}) \leq 6 \sum_{k=n}^{n+2p} OG_k.$$

Since $\sum_{n \in \mathbb{N}} OG_n$ converges, the sequence $(A_{2n})_{n \in \mathbb{N}}$ is a Cauchy sequence and converges to a point $A^\infty \in \mathcal{S}$. In the same way $(A_{2n+1})_{n \in \mathbb{N}}$ converges to A'^∞, the symmetric of A^∞ with respect to O. Moreover,

$$A_{2n} A^\infty \leq 6 \sum_{k=2n}^{\infty} OG_k \leq \frac{6}{1-q} OG_{2n},$$

for all n large enough. Therefore, $A_{2n} A^\infty = O(q^{2n})$. In the same way, we show a similar result for the other vertices of the tetrahedra. Consequently, the sequences $(\mathcal{T}_{2i})_{i \in \mathbb{N}}$ and $(\mathcal{T}_{2i+1})_{i \in \mathbb{N}}$ converge with at least geometric speed. □

Remark 27.   1. The limit is a regular tetrahedron if and only if the parameters of the non-isosceles tetrahedron $\mathcal{T}_0$ satisfy $d_{12}d_{34} = d_{13}d_{24} = d_{14}d_{23} < \frac{64}{9}$ that is $\mathcal{T}_0$ is an isodynamic tetrahedron.
  2. According to numerical experiments, we conjecture that $OG_{n+1} \sim r \times OG_n$ (convergence with order one).
  3. We use the notations of Section 2.1. The tetrahedron $\mathcal{T}_0 = (A_0 B_0 C_0 D_0)$ is not isosceles. Let $\mathcal{I}_1, \mathcal{I}_2$ be the two inversions leaving invariant the sphere $\mathcal{S}$ and mapping A_0, B_0, C_0, D_0 on the vertices of an isosceles tetrahedron (see [8, p. 184-186]). The tetrahedra $\mathcal{I}_1(\mathcal{T}_0)$ and $\mathcal{I}_2(\mathcal{T}_0)$ are isometric to the limits of the sequences $(\mathcal{T}_{2i})_{i \in \mathbb{N}}$ and $(\mathcal{T}_{2i+1})_{i \in \mathbb{N}}$. This can be proved by computing the lengths of the edges of $\mathcal{I}_1(\mathcal{T}_0)$ and $\mathcal{I}_2(\mathcal{T}_0)$.

6   Sequence of Cyclic Quadrilaterals

6.1   Degenerate Simplices

Let us consider the case where $\mathcal{T}_0$ is a cyclic quadrilateral. We see the following case as a degenerate case of the previous one.

Theorem 28. *Consider a convex cyclic quadrilateral $\mathcal{T}_0 = (A_0 B_0 C_0 D_0)$ with circumcircle $\mathcal{C}$ of center O and radius 1, and such that its vertices are pairwise distinct. We use the previous iteration to build a sequence of convex quadrilaterals $(\mathcal{T}_i)_{i \in \mathbb{N}}$. The sequences $(\mathcal{T}_{2i})_{i \in \mathbb{N}}$ and $(\mathcal{T}_{2i+1})_{i \in \mathbb{N}}$ are well-defined and converge to rectangles that have same image, whose centroid is O and whose lengths of the edges are $d_{13}^\infty = 4$, $d_{12}^\infty = 4\dfrac{\sqrt{d_{12}d_{34}}}{\sqrt{d_{13}d_{24}}}$ and $d_{14}^\infty = 4 - d_{12}^\infty$. If the limit is not a square then $OG_{n+1} \sim \dfrac{|d_{12}^\infty - 2|}{2} OG_n$ and the sequence $(\mathcal{T}_{2i})_{i \in \mathbb{N}}$ converges with at least geometric speed. In the case where the limit is a square, the speed of convergence of $(OG_n)_{n \in \mathbb{N}}$ is at least with order $\dfrac{3}{2}$.*

We keep the previous notations. In particular, for all $n \in \mathbb{N}$, the parameters $(d_{ij}^n)_{1 \leq i < j \leq 4}$ refer to the square of the lengths of the edges (diagonals included) of $\mathcal{T}_n$.

Proof. The beginning of the proof is similar to the proof of Theorem 13. By Theorem 12.2., for every $(i, j, k, l) \in \mathcal{U}$, the sequence $(d_{ij}^n d_{kl}^n)_{n \in \mathbb{N}}$ is increasing and for $1 \leq i < j \leq 4$, the sequence $(d_{ij}^n)_{n \in \mathbb{N}}$ has a positive lower bound. Thus a cluster point has pairwise distinct vertices. With Theorem 12.1., we prove that a cluster point $\mathcal{T}$ is a rectangle. The parameters d_{ij}^{∞}'s $(1 \leq i < j \leq 4)$ of $\mathcal{T}$ are given by Theorem 13.2. Since the quadrilaterals are convex, we have $d_{13}^{\infty} = 4$ and $d_{14}^{\infty} = 4 - d_{12}^{\infty}$. We put $\mu = d_{12}^{\infty}$.

For every $n \in \mathbb{N}$, there exist three relations between the h_{ij}^n's, $(1 \leq i < j \leq 4)$:

$$\Gamma(A_n, B_n, C_n, D_n) = 0, \tag{4}$$

$$a_n{}^2 b_n{}^2 c_n{}^2 = 16 \times (area(A_n B_n C_n))^2, \tag{5}$$

$$b_n'{}^2 c_n'{}^2 a_n{}^2 = 16 \times (area(B_n C_n D_n))^2. \tag{6}$$

The last two relations express that the triangles $(A_n B_n C_n)$ and $(B_n C_n D_n)$ admit a circumscribed circle with radius 1 (the value of the area is given in Relation (1)).

Part I.
The computation of the terms of degree 1 of the Taylor series of the Relation (4) gives

$$h_{12}^n + h_{14}^n + h_{23}^n + h_{34}^n - h_{24}^n - h_{13}^n = O\left(||h_n||^2\right).$$

The computation of the terms of degree 1 of the Taylor series of Relations (5) and (6) gives

$$h_{13}^n = O\left(||h_n||^2\right) \text{ and } h_{24}^n = O\left(||h_n||^2\right).$$

Thus $\delta_n = O\left(||h_n||^2\right)$.

Following the proof of Lemma 23, we see that $h_{12}^n + h_{34}^n = O\left(||h_n||^2\right)$ and $h_{14}^n + h_{23}^n = O\left(||h_n||^2\right)$.

Thus $h_{12}^n{}^2 + h_{14}^n{}^2 = \Theta\left(||h_n||^2\right)$.

Part II.
The computation of the terms of degree at most two of the Taylor series of Relations (5) and (6) gives:

$$h_{13}^n \mu(4 - \mu) + (h_{12}^n + h_{23}^n)^2 = O\left(||h_n||^3\right) \text{ and}$$
$$h_{24}^n \mu(4 - \mu) + (h_{23}^n + h_{34}^n)^2 = O\left(||h_n||^3\right),$$

that is

$$h_{13}^n = \frac{-(h_{12}^n - h_{14}^n)^2}{\mu(4-\mu)} + O\left(||h_n||^3\right) \text{ and}$$

$$h_{24}^n = \frac{-(h_{12}^n + h_{14}^n)^2}{\mu(4-\mu)} + O\left(||h_n||^3\right).$$

In the same way, Relation (4) gives:

$$\mu(\mu-4)(h_{12}^n + h_{14}^n + h_{23}^n + h_{34}^n - h_{24}^n - h_{13}^n) + \mu h_{14}^n{}^2 + (4-\mu)h_{12}^n{}^2 = O\left(||h_n||^3\right).$$

We deduce:

$$\delta_n = \frac{-h_{14}^n{}^2}{\mu} - \frac{h_{12}^n{}^2}{4-\mu} + O\left(||h_n||^3\right) \text{ and } \delta_n = \Theta\left(||h_n||^2\right).$$

Moreover,

$$\epsilon_n = 4\mu h_{12}^n{}^2 + 4(4-\mu)h_{14}^n{}^2 + O\left(||h_n||^3\right).$$

Hence,

$$OG_{n+1}{}^2 = \frac{(\mu-2)^2}{64}\left(\frac{h_{14}^n{}^2}{\mu} + \frac{h_{12}^n{}^2}{4-\mu}\right) + O\left(||h_n||^3\right),$$

$$= \frac{(\mu-2)^2}{4}OG_n{}^2 + O\left(||h_n||^3\right).$$

Case 1: $\mu \neq 2$. The limit is not a square and $OG_{n+1} \sim \frac{|\mu-2|}{2}OG_n$. The convergence is with order one.

Case 2: $\mu = 2$. The limit is a square and $OG_{n+1} = O\left(||h_n||^{\frac{3}{2}}\right)$, that is

$$OG_{n+1} = O\left(OG_n^{\frac{3}{2}}\right).$$

In both cases, the series $\sum_{n \in \mathbb{N}} OG_n$ converges with at least geometric speed. Thus we may reason as in Theorem 26 and we obtain that the sequences $(\mathcal{T}_{2i})_{i \in \mathbb{N}}$ and $(\mathcal{T}_{2i+1})_{i \in \mathbb{N}}$ converge to two rectangles, that have same image, with at least a geometric speed. □

6.2   A Particular Case

Let N be the kernel of the symmetric matrix Σ. $h_n \in N$ if and only if $h_{13}^n = h_{24}^n$.

Definition 29. *We say that a vector $h_n \in \mathbb{R}^6$ has a "good" value if and only if the $d_{ij}^{n}{}'s = (d_{ij}^{\infty} + h_{ij}^n)'s, (1 \leq i < j \leq 4)$ satisfy the relations (4),(5),(6) in Section 6.1.*

Assume that h_n has a good value and a small norm. Then a geometric argument or an algebraic computation shows that

$$h_n' \in N \Leftrightarrow \mathcal{T}_n \text{ is an isosceles trapezoid.}$$

Proposition 30. *Keeping the assumptions of Theorem 28, we assume that there exists $k \in \mathbb{N}$ such that $\mathcal{T}_k$ is an isosceles trapezoid but not a rectangle. If $d_{12}^\infty = 2$ then $OG_{n+1} \sim OG_n{}^3$.*

Proof. For all $n \geq k$, $\mathcal{T}_n$ is an isosceles trapezoid that admits (OG_k) as a line of symmetry. Therefore, the sequence $(\mathcal{T}_n)_{n \in \mathbb{N}}$ cannot rotate around O. The sequence $(\mathcal{T}_n)_{n \in \mathbb{N}}$ converges to a rectangle that admits (OG_k) as a line of symmetry. We study the rate of convergence using an explicit computation. We may assume that the line of symmetry is the axis of abscissas, that the abscissa of A_n and D_n is a_n and the abscissa of B_n and C_n is b_n. Obviously, the abscissa of the centroid of $\mathcal{T}_n$ is:

$$g_n = \frac{a_n + b_n}{2}.$$

An easy computation gives:

$$a_{n+1} = -\frac{a_n^2 b_n + 2a_n b_n^2 + b_n^3 - 4a_n}{a_n^2 - 2a_n b_n - 3b_n^2 + 4}$$

and

$$b_{n+1} = -\frac{b_n^2 a_n + 2b_n a_n^2 + a_n^3 - 4b_n}{b_n^2 - 2b_n a_n - 3a_n^2 + 4}.$$

We know that the sequence $(a_n + b_n)_{n \in \mathbb{N}}$ converges to 0 and the sequence $(a_n^2)_{n \in \mathbb{N}}$ converges to $\dfrac{d_{12}^\infty}{4}$. Hence,

$$g_{n+1} \sim \frac{(a_n + b_n)(a_n^4 - 4a_n^3 b_n - 10a_n^2 b_n^2 + 16a_n^2 - 4a_n b_n^3 + 16b_n^2 - 16 + b_n^4)}{-32}.$$

1. Case $d_{12}^\infty \neq 2$. The limit is not a square and we obtain $g_{n+1} \sim \left(1 - \dfrac{d_{12}^\infty}{2}\right) g_n$ (convergence with order one). If $d_{12}^\infty > 2$ then, for all sufficiently large n, $O \in]G_n G_{n+1}[$.

2. Case $d_{12}^\infty = 2$. The limit is a square, that is the parameters of $\mathcal{T}_0$ satisfy the relation:

$$(d_{12})^2 = d_{14} d_{23}.$$

Define two sequences $(u_n)_{n \in \mathbb{N}}$ and $(v_n)_{n \in \mathbb{N}}$ by the equalities:

$$\text{for all } n \in \mathbb{N}, \ a_n = \frac{1}{\sqrt{2}} + u_n \text{ and } b_n = -\frac{1}{\sqrt{2}} + v_n.$$

The sequences $(u_n)_{n \in \mathbb{N}}$ and $(v_n)_{n \in \mathbb{N}}$ converge to 0 and

$$\text{for all } n \in \mathbb{N}, \ (u_n, v_n) \neq (0, 0).$$

We compute the Taylor series of the previous relation and we consider the terms of degree at most 2. We obtain

$$16\sqrt{2}(u_n - v_n) = -8(3u_n^2 + 3v_n^2 + 2u_n v_n) + O\left(\|(u_n, v_n)\|^3\right).$$

Therefore,

$$u_n - v_n \sim \frac{-1}{2\sqrt{2}}(3u_n^2 + 3v_n^2 + 2u_n v_n),$$

$$u_n - v_n = O(\|(u_n, v_n)\|^2) \text{ and then}$$

$$u_n + v_n = \Theta(\|(u_n, v_n)\|).$$

We deduce easily the estimate:

$$g_{n+1} \sim \frac{(a+b)(-4(u_n + v_n)^2 + O\left(\|(u_n, v_n)\|^3\right)}{-32} \sim \frac{-4(u_n + v_n)^3}{-32},$$

that is $g_{n+1} \sim g_n{}^3$ (convergence with order three).   □

For instance if $a_0 = 0.955, b_0 = 0.12237784429$, we obtain almost a square after three iterations.

6.3   About the Limit

As in the case of a tetrahedron, there exist two inversions leaving $\mathcal{C}$ invariant and mapping A_0, B_0, C_0, D_0 on the vertices of a rectangle. The images of the cyclic quadrilateral $\mathcal{T}_0$ by these inversions are isometric to the limits of the sequences $(\mathcal{T}_{2i})_{i \in \mathbb{N}}$ and $(\mathcal{T}_{2i+1})_{i \in \mathbb{N}}$.

Remark 31.   1. The limit is a square if and only if the parameters of the non-rectangular quadrilateral $\mathcal{T}_0$ satisfy $d_{12}d_{34} = d_{14}d_{23} < 4$, that is $\mathcal{T}_0$ is a *harmonic (or isodynamic)* quadrilateral.
2. If the limit is a square then the convergence seems to be with order three. More precisely we conjecture that $OG_{n+1} \sim OG_n{}^3$.

7   Solution of the Case $d = 2$

From now on, $\mathcal{T}_0$ is a non-flat triangle. We consider a dynamical system of triangles $(\mathcal{T}_n)_{n \in \mathbb{N}}$.

Remark 32.   1. The application ϕ is not one to one: indeed, if $\mathcal{T}_1$ is a generic triangle, then there exist two triangles $\mathcal{T}_0$ such that $\phi(\mathcal{T}_0) = \mathcal{T}_1$.
2. The triangle $\mathcal{T}_1$ is said to be the circum-medial triangle associated to $\mathcal{T}_0$ (see [5, p. 162-163]).

7.1   The Parameters

Notations

1. For all $n \in \mathbb{N}$, let $a_n = B_n C_n$, $b_n = B_n C_n$, $c_n = C_n A_n$.
2. We denote the lengths a_0, b_0, c_0 by a, b, c.

Definition 33.   *1. For all $n \in \mathbb{N}$, the parameters of $\mathcal{T}_n$ are defined by:*
$$s_n = a_n{}^2 + b_n{}^2 + c_n{}^2, \quad t_n = a_n{}^2 b_n{}^2 + b_n{}^2 c_n{}^2 + c_n{}^2 a_n{}^2, \quad u_n = a_n{}^2 b_n{}^2 c_n{}^2.$$
2. We denote the parameters of $\mathcal{T}_0$ by s, t and u.

Recall that the circumradius of $\mathcal{T}_0$ is 1. Then $u = 4t - s^2$ and $t > \dfrac{s^2}{4}$. One has $u = 16 \times S^2$ where S is the area of $(A_0 B_0 C_0)$ and $OG_0{}^2 = 1 - \dfrac{s}{9}$ with $0 < s \le 9$.

Therefore,

$$s = 9 \Leftrightarrow G_0 = O \Leftrightarrow (A_0 B_0 C_0) \text{ is an equilateral triangle.}$$

We have the equality:

$$s^2 - 3t = a^4 - a^2(b^2 + c^2) + b^4 + c^4 - b^2 c^2.$$

Its RHS is a polynomial in a^2 with discriminant $-3(b^2 - c^2)^2 \le 0$. Thus $t \le \dfrac{s^2}{3}$ and

$$t = \frac{s^2}{3} \Leftrightarrow (A_0 B_0 C_0) \text{ is equilateral.}$$

7.2   Deformation from $\mathcal{T}_0$ to $\mathcal{T}_1$

Here A_0, B_0, C_0 can be on a line but are not all equal. Thus $s > 0, t > 0$.

Proposition 34. *We have $s_1 \ge s$, $u_1 \ge u$ and if $s \ge 3$, $t_1 \ge t$. Moreover the equality $s_1 = s$ holds if and only if (A_0, B_0, C_0) is an equilateral triangle.*

Proof. We obtain:

$$s_1 = \frac{s^2(6t - s^2)}{D}, \quad t_1 = \frac{s^4 t(9t - 2s^2)}{D^2}, \quad u_1 = 4t_1 - s_1{}^2,$$

where $D = -4s^3 + 18st - 108t + 27s^2$. Clearly $D > 0$ and D is bounded.

We have $s_1 - s = \dfrac{3s(9 - s)(4t - s^2)}{D} \ge 0$. Clearly, $s_1 = s$ if and only if $s = 9$ or $u = 0$, that is $A_0 B_0 C_0$ is equilateral or flat.

From the equality

$$t_1 - t = \frac{9t\left(4t - s^2\right)\left(9\left(\dfrac{s^2}{3} - t\right)(s - 6)^2 + s^2(9 - s)(s - 3)\right)}{D^2},$$

we deduce that for $s \ge 3$ the inequality $t_1 \ge t$ holds.

One has $\dfrac{u_1}{u} = \left(\dfrac{s^3}{D}\right)^2$. Hence,

$$u_1 \ge u \Leftrightarrow s^3 \ge D \Leftrightarrow v = s^2(9 - s) + 18\left(\frac{s^2}{3} - t\right)(s - 6) \ge 0.$$

If $s \ge 6$ then $v \ge 0$. For $s < 6$, $t > \dfrac{s^2}{4}$ implies that $v \ge \dfrac{s^3}{2}$.   □

7.3 Convergence of the Triangles

We assume that T_0 is not an equilateral triangle.

Proposition 35. *Let T be a cluster point of the bounded sequence $(T_n)_{n \in \mathbb{N}}$. Then T is a non-flat equilateral triangle. Moreover, the lengths of the edges of T_n converge to $\sqrt{3}$.*

Proof. 1. The non-decreasing sequence $(u_n)_{n \in \mathbb{N}}$ converges to a real $u^\infty > 0$. The increasing sequence $(s_n)_{n \in \mathbb{N}}$ converges to a real $s^\infty > 0$. Let T be a cluster point of the sequence $(T_n)_{n \in \mathbb{N}}$. With a similar proof to this one used in Theorem 26, we show that:
  (a) s^∞, u^∞ are parameters of T. Therefore T is not flat.
  (b) T is an equilateral triangle, $s^\infty = 9$ and $u^\infty = 27$.
  2. Since for all sufficiently large n, $s_n \geq 3$, the sequence $(t_n)_{n \in \mathbb{N}}$ is non-decreasing. Hence, $(t_n)_{n \in \mathbb{N}}$ converge to 27, the corresponding parameter of T. The sequences $(s_n)_{n \in \mathbb{N}}, (t_n)_{n \in \mathbb{N}}, (u_n)_{n \in \mathbb{N}}$ converge to $9, 27, 27$. Therefore, the sequences $(a_n)_{n \in \mathbb{N}}, (b_n)_{n \in \mathbb{N}}, (c_n)_{n \in \mathbb{N}}$ converge to $\sqrt{3}$. $\square$

Let $a_n{}^2 = 3 + h_n, b_n{}^2 = 3 + k_n, c_n{}^2 = 3 + l_n$ and $\delta_n = (h_n, k_n, l_n)$.

Lemma 36. *One has $h_n + k_n + l_n \sim \frac{1}{3}(h_n k_n + k_n l_n + l_n h_n) \sim \frac{-1}{6}||\delta_n||^2$.*

Proof. Obviously,

$$u = t - 4s^2 \Leftrightarrow 3(h_n + k_n + l_n) = -(h_n + k_n + l_n)^2 + (h_n k_n + k_n l_n + l_n h_n) - h_n k_n l_n.$$

Thus

$$3(h_n + k_n + l_n) = -(h_n + k_n + l_n)^2 + (h_n k_n + k_n l_n + l_n h_n) + O\left(||\delta_n||^3\right),$$
$$= h_n k_n + k_n l_n + l_n h_n + O\left(||\delta_n||^3\right).$$

Consequently, $h_n k_n + k_n l_n + l_n h_n = \dfrac{1}{2}(h_n + k_n + l_n)^2 - \dfrac{1}{2}(h_n^2 + k_n^2 + l_n^2)$
$$\sim -\frac{1}{2}(h_n^2 + k_n^2 + l_n^2).$$ $\square$

Proposition 37. *One has $OG_{n+1} \sim OG_n{}^2$.*

Remark 38. The sequence $(OG_n)_{n \in \mathbb{N}}$ converges to 0 with order 2.

Proof. From the equalities $OG_n{}^2 = \dfrac{h_n + k_n + l_n}{-9}$ and $OG_{n+1}{}^2 = 1 - \dfrac{s_{n+1}}{9}$, using Maple and from Lemma 22, we obtain the Taylor series of $OG_{n+1}{}^2$ with the precision $O\left(||\delta_n||^5\right)$:
$$OG_{n+1}{}^2 = \frac{N_n}{-81^2}, \text{ where:}$$

$$N_n = 81(h_n + k_n + l_n)^2 + 18(h_n + k_n + l_n)(2h_n{}^2 + 2k_n{}^2 + 2l_n{}^2 +$$
$$h_n k_n + k_n l_n + l_n h_n) + O\left(||\delta_n||^5\right),$$

$$= 27(h_n + k_n + l_n)(3(h_n + k_n + l_n) - 2(h_n k_n + k_n l_n + l_n h_n)) + O\left(||\delta_n||^5\right),$$
$$\sim -81(h_n + k_n + l_n)^2.$$

Finally, $OG_{n+1}{}^2 \sim \dfrac{(h_n + k_n + l_n)^2}{81} = OG_n{}^4.$ □

Remark 39. 1. We can deduce that there exists $\lambda \in (0, 1)$, that depends on a, b, c, such that $OG_n \sim \lambda^{2^n}$. If $\mathcal{T}_0$ is close to a flat triangle then λ is close to 1. Of course, if $\mathcal{T}_0$ is close to an equilateral triangle then λ is close to 0 (see $\mathcal{T}_0$ as the result of a large number of iterations.)
2. The result $OG_{n+1} = O(OG_n{}^2)$ obtained in [2] is weaker and does not give the previous estimate of OG_n.

7.4   The Main Result in Dimension Two

Theorem 40. *The sequences $(\mathcal{T}_{2i})_{i\in\mathbb{N}}$ and $(\mathcal{T}_{2i+1})_{i\in\mathbb{N}}$ are well-defined and converge with at least quadratic speed to two equilateral triangles that are symmetric with respect to O.*

Proof. As in the proof of Theorem 26, we show that the sequence $(A_{2n})_{n\in\mathbb{N}}$ converges to a point $A^\infty \in \mathcal{C}$ and that $(A_{2n+1})_{n\in\mathbb{N}}$ converges to a point A'^∞, the symmetric of A^∞ with respect to O. Moreover,

$$A_{2n} A^\infty \leq 6 \sum_{k=2n}^{\infty} OG_k \leq 12 \times OG_{2n}$$

for all n large enough. Therefore, $A_{2n} A^\infty = O\left(\lambda^{2^{2n}}\right)$. In the same way, we show a similar result for the other vertices of the triangles. Consequently, the sequences $(\mathcal{T}_{2i})_{i\in\mathbb{N}}$ and $(\mathcal{T}_{2i+1})_{i\in\mathbb{N}}$ converge with at least quadratic speed.   □

8   Conclusion

We mention two open problems about these dynamical systems:

1. It is a natural question to wonder whether this result can be generalized to the case $d \geq 4$. That is, does the sequence of d-simplices $(\mathcal{T}_{2i})_{i\in\mathbb{N}}$ converge to a d-simplex with centroid O ? Unfortunately, by the methods of our paper, the complexity of the computation increases considerably with d.
2. We return to the case $d = 3$. It seems to us interesting to replace the centroid G_i of $\mathcal{T}_i$ with some other center Ω_i of $\mathcal{T}_i$. We may choose, for instance, Ω_i as the incenter, the Monge point or the Fermat-Torricelli point of $\mathcal{T}_i$. Indeed if one among these centers and the circumcenter of a tetrahedron $\mathcal{T}$ coincide, then $\mathcal{T}$ is isosceles (see [4, p. 494]). Now let us suppose that we can prove that the sequence $(\Omega_{2i})_{i\in\mathbb{N}}$ converges to O with at least a geometric speed. Then we can deduce, using the previous methods, that the sequence $(\mathcal{T}_{2i})_{i\in\mathbb{N}}$ converges with at least a geometric speed to an isosceles tetrahedron.

Acknowledgments. The authors wish to thank the referees for their helpful comments and D. Adam for many valuable discussions.

References

1. Altshiller-Court, N.: Modern Pure Solid Geometry, 2nd edn., p. 353. Chelsea Publishing Co., New York (1979)
2. Bourgeois, G., Lechêne, J.P.: Etude d'une itération en géométrie du triangle. Bulletin APMEP 409, 147–154 (1997) (in French)
3. Coxeter, H., Greitzer, S.: Geometry Revisited. The Mathematical Association of America (1967)
4. Edmonds, A.L., Hajja, M., Martini, H.: Coincidences of Simplex Centers and Related Facial Structures. Contributions to Algebra and Geometry 46(2), 491–512 (2005)
5. Kimberling, C.: Triangle Centers and Central Triangles. Congressus Numerantium, vol. 129. Utilitas Mathematica, Winnipeg (1998)
6. Lisoněk, P., Israel, R.B.: Metric Invariants of Tetrahedra via Polynomial Elimination. In: Proceedings of the 2000 International Symposium on Symbolic and Algebraic Computation (St. Andrews), pp. 217–219 (electronic). ACM Press, New York (2000)
7. Mitrinović, D.S., Pečarić, J.E., Volenec, V.: Recent Advances in Geometric Inequalities. Mathematics and its Applications (East European Series), vol. 28, p. 710. Kluwer Academic Publishers Group, Dordrecht (1989)
8. Thébault, V.: Parmi les plus belles figures de la géométrie dans l'espace. Géométrie du tétraèdre. Vuibert (1955) (in French)
9. Yang, L.: Solving Geometric Constraints with Distance-Based Global Coordinate System. In: International Workship on Geometric Constraint Solving, Beijing (2003)

On the Design and Implementation of a Geometric Knowledge Base

Xiaoyu Chen[1,*], Ying Huang[1], and Dongming Wang[2]

[1] LMIB – SKLSDE – School of Mathematics and Systems Science,
Beihang University, Beijing 100191, China
[2] Laboratoire d'Informatique de Paris 6, Université Pierre et Marie Curie – CNRS,
4 place Jussieu – BP 169, 75252 Paris cedex 05, France

Abstract. This paper presents the design of a geometric knowledge base that stores standardized, formalized, and structured geometric knowledge data. Emphasis is placed on the classification and organization of such knowledge data. In order to master the complexity of data relations, we adopt a key strategy that works by encapsulating certain interrelated knowledge data into knowledge objects and then organizing the knowledge objects according to the hierarchic structure of their relations. We also present our preliminary implementation of a geometric knowledge base system that provides functionalities for creating, rendering, and managing knowledge data with basic query services.

Keywords: geometric knowledge, knowledge data, knowledge object, knowledge base, knowledge management.

1 Introduction

1.1 Motivation

Geometry, a subject of mathematics that models objects and relations abstracted from the real visual world, involves knowledge about both abstract quantities and intuitive figures. Quantities and figures are interdependent in the study of geometry, in particular for solving problems involving complicated calculations (with coordinates and algebraic quantities as in analytic geometry), performing automated reasoning (for which algebraic methods are the most powerful), and rendering figures on computer screen (which are located in the coordinate systems within the screen). Therefore, geometric knowledge is not bounded within geometry itself, but also involves algebraic elements. It is geometric description, its algebraic counterpart, and diagrammatic presentation together that constitute the entirety of geometric knowledge. For example, to depict a circle, we may need a statement "a circle with point O as its center and r as its radius", an algebraic equation "$(x - a)^2 + (y - b)^2 = r^2$", where (a, b) is the pair of coordinates of point O, and a diagram of the circle on the screen.

---

* This paper was prepared when the author was visiting the DAM group of Department of Mathematics and Computer Science, Technische Universiteit Eindhoven.

T. Sturm and C. Zengler (Eds.): ADG 2008, LNAI 6301, pp. 22–41, 2011.

In general, the following three kinds of representations are needed for geometric knowledge.

- Declarative representation: explicit, formal and natural language statements to represent definitions of geometric concepts, theorems of geometric theory, proofs of geometric theorems, etc.
- Procedural representation: specifications of algorithms, methods, rules, or heuristics to calculate the values of geometric quantities, to translate geometric configurations into algebraic relations between coordinates, to solve geometric problems or their algebraic counterparts, etc.
- Diagrammatic representation: diagram instructions to construct or visualize geometric configurations.

In recent years, many efforts have been devoted to the development of high-level software for dynamic geometry, mechanical geometric reasoning, and symbolic geometric computation. Each piece of such software has its own implementation of geometric knowledge for the representation and algebraization of geometric objects and relations. However, many of the knowledge data used in these different pieces of geometry software are similar or even the same semantically. For example, a circle may be constructed by three points, two points, or a point and a segment in almost every dynamic geometry software system. It is time-consuming and effort-wasting for one to specify such sophisticated knowledge again while developing new geometry software. It is thus helpful to build up a repository of geometric knowledge as explicit standard specifications for implementing some basic functionalities of geometry software.

On the other hand, geometric knowledge is accumulated step by step, e.g., by introducing new concepts using already defined concepts, deriving useful properties about new concepts, and proving or discovering theorems relating old and new concepts. It does not lie at a flat level but is piled up with certain intrinsic structure of hierarchy. To study and capture the knowledge structure (which is the meta-knowledge about geometric knowledge), we note the following two aspects.

(1) For education, geometric knowledge need be distributed to learners in the style of documents such as textbooks, lecture notes, and papers. There are common practice and implicit convention in the scientific community as how knowledge should be organized, formulated, and presented. Some basic knowledge serves as preliminaries for high-level knowledge. The narrative structure of a document depends on the logical structure of the knowledge involved, e.g., from the simplest to the most complicated and from the basic to the advanced. Knowledge structure may be acquired and used to assist authors to create sound documents [2].

(2) For research, geometric knowledge need be structured so that it is easy to trace how knowledge is developed and to identify what knowledge is basic and what knowledge is derived. Geometric configurations can be described in different ways by using different geometric concepts. For instance, the perpendicular foot of two lines can also be described as the intersection point of two perpendicular lines. This kind of variants of descriptions for geometric configurations makes it difficult to retrieve knowledge data semantically. It is necessary to study knowledge structure so that geometric configurations can be represented in

(or translated according to the definitions of the involved concepts into) a canonical or standard form.

Geometric knowledge bases are created to store geometric knowledge data, with well-defined structures for the types of and relationships between the knowledge data stored. They should provide standard specifications for implementing basic functionalities needed in automated deduction, diagram generation, problem solving etc., and friendly interfaces for sharing and reusing the knowledge data created by users to generate human-readable geometric documents.

The knowledge bases should also support manipulations such as browsing, creating, removing, and modifying knowledge data and queries for retrieving pieces of knowledge indexed by keywords or related to given pieces of knowledge to assist authors to create electronic documents.

The creation of a geometric knowledge base consists in providing data and support for the development of geometry software (such as specifications of geometric concepts for describing geometric configurations, rules for the translation of geometric concepts into algebraic expressions, instructions for the construction of diagrams for complicated concepts, e.g., the Napoleon triangle of a triangle), and facilities for the management of structures of geometric knowledge. The availability of geometric knowledge bases may help reduce individual efforts on creating knowledge data for geometry software development, geometric document authoring, and other geometric applications.

1.2   State of the Art

Several projects have been initiated for the purposes of standardizing representations of geometric configurations and setting up a publicly available repository for share and reuse.

GEOTHER [10] is an Epsilon module implemented as an integrated environment in Maple for translating the predicate specification of a geometric theorem into English or Chinese statement, first-order logical formula, or algebraic expressions, automated proving by algebraic methods, generating dynamic diagrams, and documenting the results. It provides a standard form for specifying the entries contained in the predicate routines which define how to manipulate and translate the predicates at the running time. However, the predicates only play the role of function calls whose entries are not complete in the sense of geometric knowledge, since the definitions of the predicates are not provided. Moreover, the relations among the defined predicates and specified geometric theorems are not taken into account.

GeoCode [4] is a generic proof scheme language standard that can be run by GeoProver which translates the proof scheme of a geometric theorem into the specific form of the target proving system and implements the interfaces for different CAS to solve the algebraic counterpart of the geometric problem by algebraic methods. To specify and manipulate geometric theorems, the standard provides the routine code (translating into algebraic expressions and construction steps for drawing the diagrams) of basic functions and advanced functions that are defined by using basic functions. However, the structure of these functions and the collected proof schemes is not studied.

GeoThms [8] is an Internet framework that integrates dynamic geometry software (GCLC, Eukleides), geometry theorem provers (GCLCprover), and a repository storing geometric problems, their statements, illustrations, and proofs. Although this framework provides a platform for users to browse and learn geometric theorems, explore new geometric conjectures specified in the specific language, and obtain the automatically generated results, some geometric knowledge, such as the definitions of geometric concepts and the rules for translating geometric constructions into their algebraic counterparts, is not involved. Moreover, the relations among theorems, such as what theorems can be grouped together and what theorems can be used in the proof of a theorem, are not taken into account.

Intergeo [5] is an ongoing European project, whose objectives are to attack the barrier of lack of interoperability by offering a common file format for specifying dynamic diagrams and to build up an annotation and search web platform which provides access to thousands of interactive geometry constructions by crossing the boundaries of the curriculum standards of Europe. The project defines an ontology used in searching, which specifies the cross-curriculum characterization of geometric constructions. The platform shares the learning resources on the form of interactive geometric constructions and related materials, which are different from our targets of investigation.

The main novelty of our work lies in (1) encapsulating interrelated data of geometric knowledge presented in the literature and used in geometry software into *knowledge objects*, (2) formalizing the encapsulated geometric knowledge objects for representation in a knowledge base, and (3) highlighting the importance of analyzing and structuring the relations among geometric knowledge objects. With the data and objects stored in the knowledge base, functionalities may be (easily) implemented to translate formalized geometric configurations into algebraic representations, into statements in natural languages, and into diagram constructions automatically and to support the creation of dynamic geometric documents.

1.3   Problem

In [2], two of the authors proposed to develop a system enabling users to construct and manage dynamic textbooks interactively. The emphasis was placed on presenting a bird's-eye view and the conception of the system and discussing the main ideas, design methodologies, and involved tasks and problems that need be investigated. The system should be connected to a knowledge base capable of storing and organizing textbook knowledge data for reuse. However, design principles and implementation techniques on the creation of such a knowledge base were studied and explained only briefly.

This paper focuses on the problem of realizing a geometric knowledge base that stores standardized, formalized, and structured geometric knowledge data and addresses a number of design and implementation issues. It is required that the knowledge data stored in the database can be retrieved and used in the development of geometry software tools for automated deduction, diagram generation, and document creation and can be rendered into a natural style for human readability. The knowledge base should be updated dynamically and immediately once an update operation such as adding, removing, or modifying on the knowledge base is performed. The problems addressed are as follows.

1. Classification. Geometric knowledge involves many different types of knowledge data, such as the name and specification of a concept, the natural language statements and formal representation of a theorem, and the static or dynamic figures for a theorem. We use *data element* to denote the data unit defined in the knowledge base. Technically, data elements describe the logical unit of data and fields are the actual storage units. For the convenience of narrative, we also say that the data are *stored* in these data elements. What data elements should be abstracted to characterize geometric knowledge?
2. Representation. How to formalize and represent the data for each data element?
3. Organization. How to arrange and organize the data elements storing geometric knowledge data and what are the relationships among them?

This paper presents our ideas on assembling geometric knowledge data and mastering the complexity of relations within an amount of geometric knowledge. The first step is to define the data elements for characterizing geometric knowledge and encapsulate certain interrelated data into knowledge objects, and the knowledge base will collect an amount of these knowledge objects; this will be discussed in Sect. 2. The next step, as will be explained in Sect. 3, is to manage the structure over these knowledge objects. In Sect. 4, we shall be concerned with the implementation aspect and report our progress and experiments on a preliminary implementation of the geometric knowledge base system. The paper will be concluded with some discussions on future work. To be specific, we restrict ourselves to elementary geometry with plane Euclidean geometry as the target of investigation.

2   Representation of Geometric Knowledge Objects

In order to represent, manipulate, and manage data elements in the knowledge base, we need to identify them and to encapsulate the interrelated ones by analyzing the relationships among them.

2.1   Identification of Knowledge Data Elements

Data elements are the basic units which are defined in the knowledge base. We first identify what data elements should be included, based on an analysis of functionality requirements for the knowledge as follows.

– Geometric knowledge may be rendered into natural languages, such as English and Chinese, for users to browse and read.

- Geometric problems may be represented also algebraically, so that algebraic methods (such as Wu's method and the method of Gröbner bases) can be used for automated solving.
- Geometric configurations may be visualized as static or dynamic figures.
- Nondegeneracy conditions may be associated to geometric configurations to ensure rigorousness and unambiguousness.
- Geometric knowledge may be represented in a standard formal language, so that it can be processed and manipulated by software tools for further applications (e.g., automated checking of syntax/grammar and translation of the specifications of geometric configurations into other languages or forms — natural languages, algebraic representations, and diagram drawing instructions).
- Knowledge may be retrieved through keywords.

Concepts provide terms for the description of domain knowledge. The basic components of knowledge data are specific concepts from the domain (of geometry here). The data elements for specifying geometric concepts should be provided in the knowledge base at first.

Geometric concepts can be classified into *geometric objects* (such as "point", "line", and "circle"), *geometric quantities* (such as "length", "area", and "degree"), *object relations* (such as "parallel" and "perpendicular"), and *quantity relations* (such as "equal to" and "greater than"). As geometric concepts are introduced usually in a constructed way, we represent each concept with a data element *vocabulary* and a data element *attributeList*. For example, a triangle can be represented by a vocabulary "triangle" together with an attributeList "(Point,Point,Point)" indicating that its attributes are geometric objects of type Point. The vocabulary element stores a *predicate symbol* (denoting an object relation or a quantity relation) or a *function symbol* (denoting a geometric object or quantity), which is used to distinguish one concept from others. The attributeList element stores the type constraint on the concept for the convenience of grammar checking or concept matching in applications. For example, in an instance of triangle of the form "triangle(A, B, C)", the vertices A, B, and C of the triangle must be objects of type Point. Complicated instances of concepts can be constructed by nesting instances with correct types. For example, "triangle(midpoint(A, B), midpoint(B, C), midpoint(A, C))" denotes the midpoint triangle of triangle ABC, where A, B, and C are all points.

The definitions of concepts both in a formal language for automated activities like reasoning and in natural languages for display are stored as data elements *formalDefinition* and *naturalRepresentation* respectively.

Nondegeneracy conditions for concepts to be meaningful are also stored as a data element *nondegeneracyCondition*. For example, the nondegeneracy condition for the concept "triangle" is: the three vertices of the triangle are not collinear.

Figures, both static and dynamic, are stored as data elements *staticFigure* and *dynamicFigure* for visualizing the corresponding configurations.

Scripts for translating concepts into natural languages, algebraic representations, and instructions for drawing diagrams are stored as data elements *translationScript*, *algebraicScript*, and *diagramScript* respectively.

Since the data elements mentioned above are all associated with Concept, we encapsulate their data together with their relations into a *geometric concept object* (called *predicate* or *function with embedded knowledge* in [11]).

More generally, we refer to a set of structured and itemized geometric knowledge data as a *geometric knowledge object* (*KO* for short). Geometric knowledge objects can be classified into Concept (Definition), Axiom, Lemma, Theorem, Corollary, Conjecture, Proof, Problem, Example, Exercise, Solution, and Algorithm. Although this classification may be arguable and need to be justified, what is essential in our approach is to encapsulate interrelated data into certain knowledge objects with the same structure.

The objects of Lemma, Theorem, Corollary, Conjecture, Problem, Example, and Exercise involve data elements

- *knowledgeName*: the data as identifiers for the knowledge objects;
- *formalRepresentation*: the data to be processed by other tools or modules for query, automated reasoning, computation, translation, etc.;
- *naturalRepresentation*: the data to be displayed for users to read;
- *diagramInstruction*: the data to be applied as instructions for drawing diagrams corresponding to the knowledge objects;
- *algebraicRepresentation*: the data made up of algebraic formulae corresponding to the knowledge objects.

The objects of Axiom, which are true assertions on the basis of beliefs or observations without formal proof, involve data elements *formalRepresentation*, *naturalRepresentation*, and *diagramInstruction*. The objects of Proof, Solution, and Algorithm involve data elements *formalRepresentation* for mechanical checking and *naturalRepresentation* for human reading. The data elements *knowledgeName* and *keyWords* need be associated with all the knowledge objects for query. The data elements *staticFigure* and *dynamicFigure* need be associated with all the knowledge objects for visualization.

2.2 Formalization of Data Elements

The data stored in data elements are not simply of type `String` or `Integer`, but have specially designed formats. Geometric knowledge presents the features and properties of geometric configurations. Geometric configurations are usually described in a way by using labels to represent objects and quantities. To formalize the descriptions of geometric configurations, we introduce the following notions.

Definition 1. The *terms* of a concept are defined as follows:

(1) a variable, such as A or B, is a term;
(2) a constant, such as 0 or π, is a term;

(3) a function of the form $f(a_1, \ldots, a_n)$, where f is the function symbol of a concept, each a_i is a term, and $(a_1, \ldots, a_n)$ satisfies the type constraint on the concept, is a term.

Definition 2. *Clauses* are defined as follows:

(1) a reference of the form $A := f(a_1, \ldots, a_n)$, where A is a variable and $f(a_1, \ldots, a_n)$ is a function, is a (reference) clause, and A is said to be *referenced* by the function $f(a_1, \ldots, a_n)$;
(2) a predicate of the form $P(t_1, \ldots, t_m)$, where P is the predicate symbol of a concept, each t_i is a term of the concept, and $(t_1, \ldots, t_m)$ satisfies the type constraint on the concept, is a (predicate) clause;
(3) if M and N are predicate clauses, then $M \wedge N$, $M \vee N$, $(M \wedge N)$, $(M \vee N)$, and $\neg M$ are also predicate clauses and thus are clauses.

A *geometric configuration* is a statement of the form $C_1; \ldots; C_k$, where each C_i is a clause and all the variables are referenced by the functions occurring in the statement. For example, a configuration for "Simson's theorem" is "A := point(); B := point(); C := point(); D := pointOn(circle(A, B, C)); J := foot(D, line(A, B)); H := foot(D, line(B, C)); G := foot(D, line(A, C))".

A *formal representation* of a theorem is a statement of the form $H \Rightarrow C$, where H is a geometric configuration, C is a predicate clause, and all the variables occurring in C are referenced by the functions occurring in H. For example, a formal representation of "Simson's theorem" is "A := point(); B := point(); C := point(); D := pointOn(circle(A, B, C)); J := foot(D, line(A, B)); H := foot(D, line(B, C)); G := foot(D, line(A, C)) $\Rightarrow$ collinear(J, H, G)".

The general structure of definitions is "concept A is defined by R which satisfies some constraints." A *formal definition* is a statement of the form $A \triangleq [R \text{ where } C]$, where A is an instance of a concept (i.e., a function or a predicate clause) to be defined, R is a term or a predicate clause, C is a predicate clause, and all the variables occurring in R but not in A are typed. For example, the intersection point of two lines l and m may be defined as "intersection$(l, m) \triangleq$ [A::Point where incident$(A, l) \wedge$ incident(A, m)]".

We give some examples for other data elements.

The *translation script* for translating the concept "collinear(Point, Point, Point)" into an English statement is "collinear(A, B, C) $\mapsto$ A, B, and C are$_v$ collinear", where are$_v$ means that "are" is a verb. The *algebraic script* for translating the concept "collinear(Point, Point, Point)" into an algebraic expression is "collinear(A, B, C) $\mapsto -$B[2]*C[1] + B[2]*A[1] + A[2]*C[1] + B[1]*C[2] $-$ B[1]*A[2] $-$ A[1]*C[2] $=_v$ 0", where P[i] stands for the ith coordinate of P. The *diagram script* for generating diagram instructions of the concept "triangle(Point, Point, Point)" is "triangle(A, B, C) $\mapsto$ segment[A, B], segment[B, C], segment[A, C]". The *nondegeneracy condition* for the concept "triangle(Point, Point, Point)" is "triangle(A, B, C) $\not\mapsto \neg$ collinear(A, B, C)".

The *natural representation* of "Simson's theorem" is "The feet of the perpendiculars from a point on the circumcircle of a triangle to the three sides of the triangle are collinear". The *diagram instruction*[1] of "Simson's theorem" is "A = $(-11.1, 1.62)$; B = $(-22.84, -15.37)$; C = $(7.14, -15.75)$; a = Polygon[A, B, C]; o = Circle[A, B, C]; D = Point[o]; l = PerpendicularLine[D, line[A, B]]; m = PerpendicularLine[D, line[B, C]]; n = PerpendicularLine[D, line[A, C]]; J = Intersect[l, line[A, B]]; H = Intersect[m, line[B, C]]; G = Intersect[n, line[A, C]]; SimsonLine = line[H, G]".

The formalization of other data elements such as the *formal representation* of proofs and exercises is under investigation.

2.3   Structure Design within Knowledge Objects

Each data element for a knowledge object may contain multiple data of the same format. For example, the definition of a concept may be represented in different

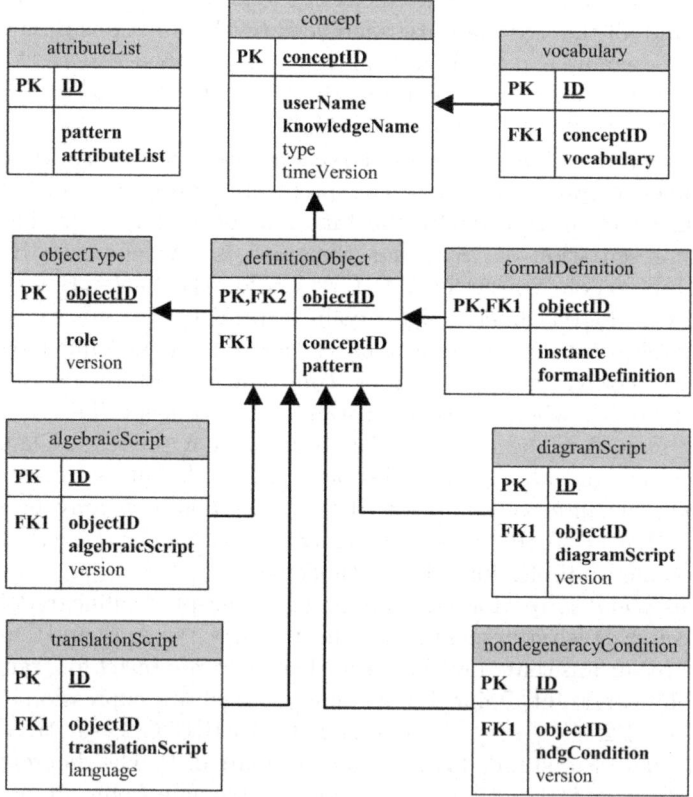

Fig. 1. Structure within geometric concept objects

---

natural languages and a theorem may have distinct algebraic representations under different coordinate systems.Therefore, it is necessary to design the structure of data elements for knowledge objects. We abstract 10 relational tables from the above analysis about the data elements for concept objects and the data model is shown in Fig. 1.

As each concept may have multiple vocabularies (defined by users) and attribute lists (e.g., a circle may be defined by three points on it or by its center and one point on it), we need construct both *vocabulary* table and *attributeList* table and use the attribute *objectID* of the *definitionObject* table to identify the same geometric concept in different situations. The attribute *conceptID* of the *concept* table is the identifier for a set of the encapsulated geometric concept objects with common concept name in the knowledge base.

Now we give an example to show how to fetch needed data from an instance of a concept according to the structure. It can be viewed as some kind of retrieval.

The expression "line(midpoint(A, B), C)" denotes the line passing through the middle point of AB and point C. The process of fetching the algebraic script

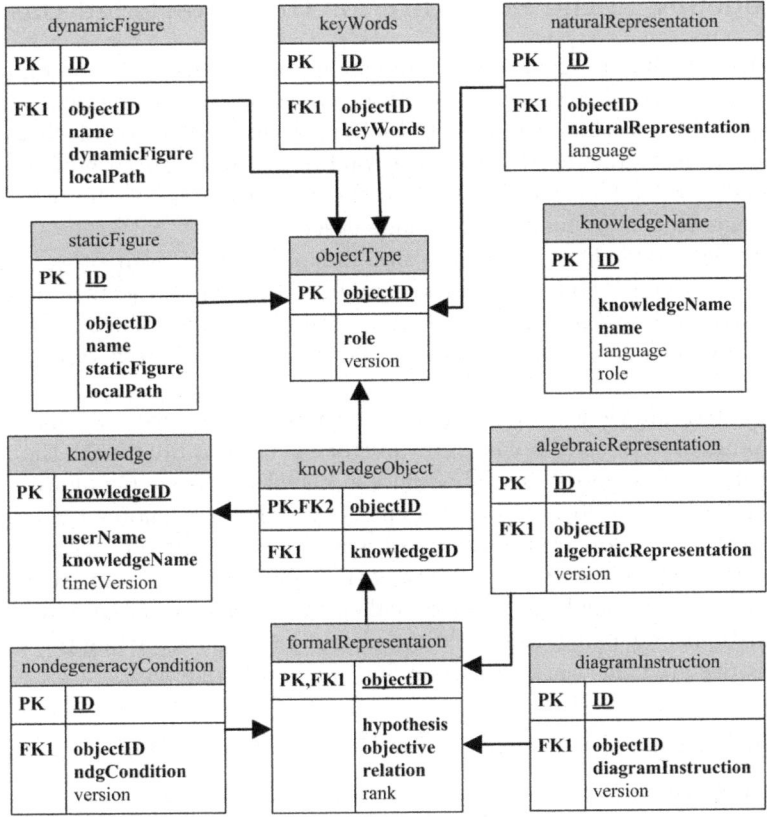

Fig. 2. Structure within geometric knowledge objects

for translating this expression into algebraic expressions is as follows. From the vocabulary "line", the conceptID "Concept.line" for the outer concept can be obtained from the vocabulary table. From the inner concept vocabulary "midpoint", the conceptID and then its type "Point" can be obtained from the vocabulary table and concept table. For the second parameter "C", the default type is "Point". From these two types, the pattern of the attribute list can be computed, e.g., by linking two "Points" into "(Point, Point)". Based on the conceptID "Concept.line" and pattern, the objectID for this expression can be obtained from the definitionObject table and further it is then feasible to fetch the needed version of algebraic script for "line(midpoint(A, B), C)" from the algebraicScript table.

Since the same data element may be used in different types of knowledge objects, it is possible to combine the structures together for other types of knowledge objects without designing separated structure for each type of knowledge objects (see Fig. 2).

According to the structure within knowledge objects, it is easy to develop tools for retrieving needed data from any given knowledge object.

3   Modeling of the Structure for the Knowledge Base

Knowledge objects are the atomic components of knowledge in communication and documentation. For example, definitions of concepts and theorems are standard knowledge units in textbooks. To build up a knowledge base of encapsulated knowledge objects, one needs to analyze and design the structure of the knowledge objects. We adopt *graph-* and *tree-based* approaches for the organization and management of geometric knowledge objects.

Geometric knowledge objects are not independent of each other; they are closely connected. Establishing relations among knowledge objects and modeling the KO structure are indispensable for designing the knowledge base. By *KO structure* we mean an arrangement of the knowledge objects based on their relations.

Interrelated knowledge objects may be grouped into *categories*, like sections and chapters used in textbooks. Here category is used to illuminate the relation between its subject and its members in the knowledge base. Categories are defined as follows: (a) $C[o_1, \ldots, o_n]$ is a category, where C denotes the subject of the category, $o_1, \ldots, o_n$ are knowledge objects, and $n \geq 1$; (b) $C'[c_1, \ldots, c_m]$ is a category, where C' denotes the subject of the category, $c_1, \ldots, c_m$ are categories (subcategories) or knowledge objects, called the *members* of the category, and $m \geq 1$. Establishing categories for knowledge objects and modeling the structure for categories are other tasks for designing the knowledge base.

3.1   Analysis on the Structure for the Knowledge Base

KO Structure. First of all, let us analyze the relations among knowledge objects, such as the concept objects and the theorem objects, which are the main components of the current version of our knowledge base. We place our emphasis on what the structure is rather than how to generate the structure.

A new concept is introduced by using some already defined concepts and it has two entries: one is used to indicate what are its father concepts and the other is used to collect some "special" constraints for the new concept. For example, the middle point of a segment is a "special" point. Its father concept is point and the special constraint is that the distances from this point to the two ends of the segment are equal. Thus, there are two kinds of KO structures that describe relations among concepts: one is *Type structure* used to describe *inheritance* relations and the other is *Derivation structure* used to describe *dependence* relations among concepts.

We have examined some concepts commonly used in elementary geometry textbooks and their Type structure is shown by the examples in Fig. 3, where the nodes represent concept objects and the directed edges represent inheritance relations. Based on our observations, ten geometry primitives have been chosen as the basic (root) types (which have no father types) for geometric objects. These primitives are defined by informal descriptions in natural mathematical languages. Although intuitively *line* is not primitive as it may be defined by means of points, from the viewpoint of geometric computation and deduction it is convenient to consider line, as well as *circle*, as primitives. For the same reason, *polygon* formed by lines or segments is also considered as primitive. The `Quantity` type consists of `Geometric Quantity` type (such as length, area, and degree) and `Algebraic Quantity` type used to define algebraic quantities (such as real numbers). The `Boolean` type consists of `Object Relation` type used to define relations among geometric objects and `Quantity Relation` type used to define relations among quantities.

The Type structure is tree-like and permits multiple inheritance. For example, an isosceles right triangle is both a special right triangle and a special isosceles triangle.

As for the Derivation structure, each concept may be derived from several concepts and it may derive many other concepts. For instance, the concept of incenter of a triangle is derived from the concepts of intersection and bisector, and it derives the concept of inscribed circle of a triangle.

Actually, the Derivation structure may also be used to describe many other dependence relations among knowledge objects: a definition may derive a theorem, some definitions may be the context of a theorem, a lemma may imply a theorem, a theorem may imply a corollary, and a proof may be associated with a theorem. For example, the definitions of "foot" and "collinear" provide the context of "Simson's theorem" and "Simson's theorem" derives the definition of "Simson line".

If we take knowledge objects as nodes and inheritance or dependence relations as directed edges, then both the Type structure and the Derivation structure can be modeled as directed graphs without directed cycles and thus we can represent these two structures in the knowledge base by the same method.

Structured with those relations among the knowledge objects, our geometric knowledge base can be completely formalized in the sense that all the non-primitive concept objects are created *formally* (and sometimes *inductively*) in

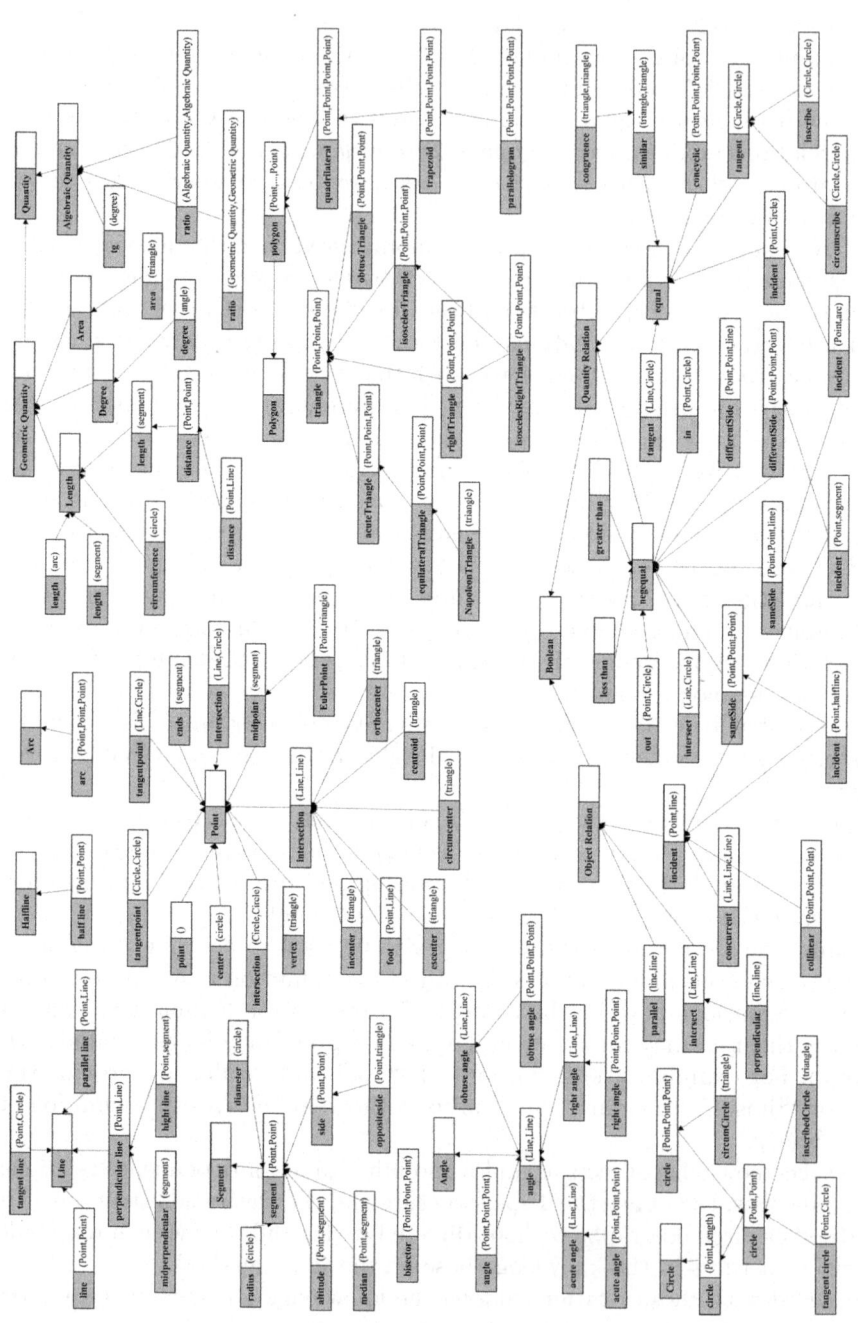

Fig. 3. Type structure

terms of other knowledge objects of ancestor types and all the knowledge objects can be traced via the relations among them.

Category. The structure of a category is a tree with subject as its root, sub-categories as its branches, and knowledge objects as its leaves. We introduce *category objects* into the knowledge base to store information about the subjects of categories. For a category object, the attribute *objectID* stores its identifer (see Fig. 4). The attribute *categoryID* stores identifiers for the variants in different natural languages. The attribute *role* represents what the category is, like chapter or section. The attribute *name* represents the name of the subject. The attributes *title* and *note* represent the contents of the subject. The relation between a category object and its member is *inclusion*.

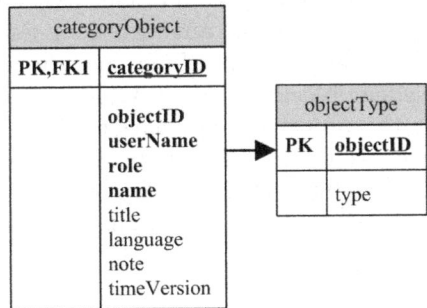

Fig. 4. Structure within the category object

3.2 Representation of the Structure of the Knowledge Base

As mentioned in Sect. 3.1, both the Type structure and the Derivation structure can be modeled as directed acyclic graphs whose nodes represent knowledge objects and whose edges represent relations between the knowledge objects. In a database the easiest way is to use a *table* whose two columns represent the starting and ending nodes connected with each other. Thus each row of the table represents a directed edge in the graph. We use a *KOstructure* table (see Fig. 5) to represent the KO structure of the knowledge base. The attributes *precursor* and *subsequence* represent the starting node and the ending node respectively, each of which stores the *objectID* of the knowledge object which the node denotes. The attribute *relationType* stores the *type* of the dependence relation between the *precursor* and the *subsequence*, such as "derive", "imply", and "context".

Similarly to the KO structure above, we use a *categoryStructure* table (see Fig. 5) to represent the structure of categories in the knowledge base. The attribute *precursor* represents a category object and stores its *objectID*. The attribute *subsequence* stores the *objectID* of its member. The attribute *rank* represents in which position in the *precursor* the *subsequence* is.

KOstructure	
PK	**ID**
	precursor subsequence relationType

categoryStructure	
PK	**ID**
	precursor subsequence rank

Fig. 5. Representation of the structure in the knowledge base

Currently, we focus our attention mainly on the inheritance relation and dependence relations among knowledge objects in order to be able to investigate what knowledge objects are basic and how to arrange knowledge objects in geometric documents, where the dependence relations indicate the narrative structure. Not only concept objects, relations among other knowledge objects, such as theorems, proofs, and solutions, are also of our concern. We have presented the structure within knowledge objects in Fig. 2. Our aim is to establish a macroscopical framework for the processing of relations among different knowledge objects (which is also why we try to encapsulate knowledge data into knowledge objects with appropriate granularity). Under this framework, we can manage the global structure and explore the inherent relationship of knowledge in geometric documents.

4   Implementation of the Knowledge Base System

We have implemented an experimental system for our geometric knowledge base. According to the structure design of data elements for knowledge objects in Sect. 2.3, we create a database containing the involved tables in MS SQL Server. We have chosen Java as the programming language to develop interfaces for users to create and modify the knowledge data of Concept (Definition), Axiom, Lemma, Theorem, Corollary, Conjecture, Problem, Example, Exercise, Proof, Solution, and Category objects and their relations and to store the constructed data into the corresponding tables in the database. Interfaces of our system with several external packages for authoring special data have been implemented. The dynamic mathematics software GeoGebra is used for creating dynamic figures. The MathDox formula editor [7] is used to create expressions in the algebraic representations, encoded in OpenMath [1], a standard for representing mathematical objects with semantics. One can construct, for example, the definition of "Simson line" as in Fig. 6 and a category of section as in Fig. 7.

4.1   Naming Objects in the Knowledge Base

In the knowledge base, knowledge objects and category objects are the units that may be managed, processed, and retrieved. A knowledge object or a category object is identified only by matching the value of its attribute *objectID*. It is

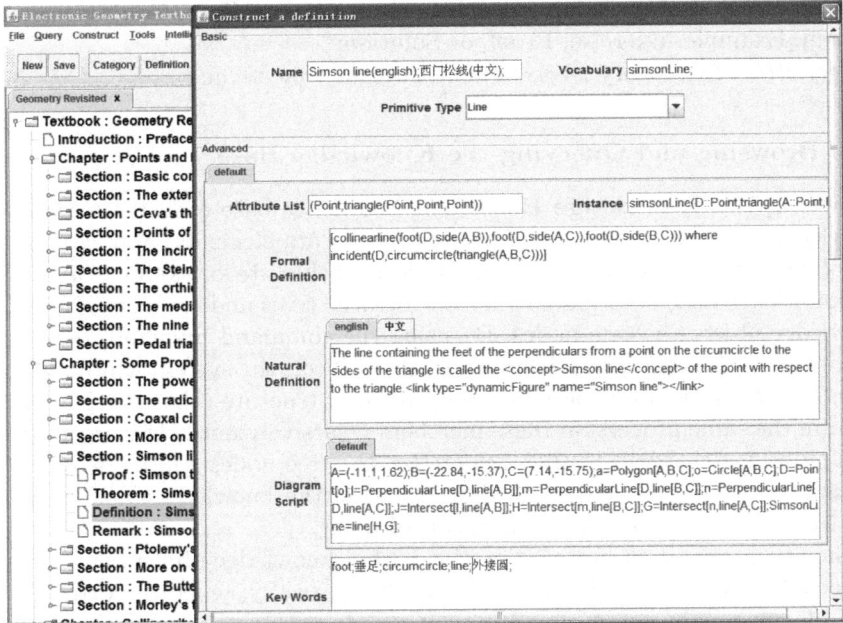

Fig. 6. Constructing the definition object "Simson line"

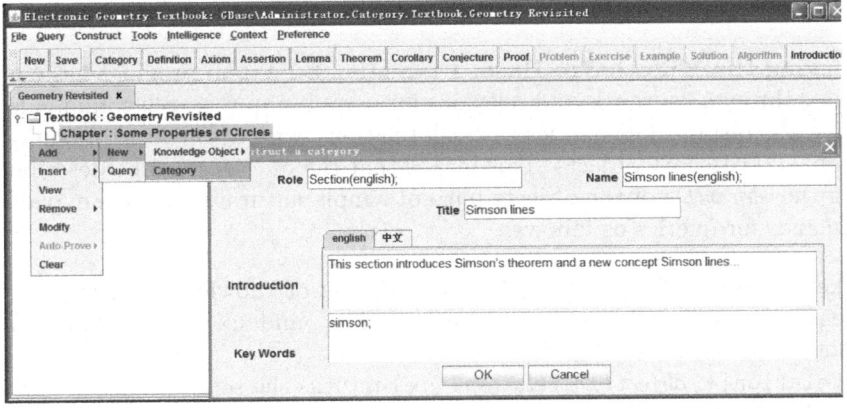

Fig. 7. Constructing the section object "Simson lines"

important to automatically generate unique *objectIDs* for each knowledge object and each category object in the knowledge base to avoid conflict. We adopt the following naming rules for generating *objectIDs* from the input data:

- *objectID* of Concept := *userName*.Definition.*knowledgeName.attributeList*;
- *objectID* of *type* := *userName.type.knowledgeName.version*

where *type* can be Axiom, Lemma, Theorem, Corollary, Conjecture, Problem, Example, Exercise, Proof, or Solution;
- *objectID* of Category := *userName*.Category.*role.name*.

4.2   Browsing and Querying the Knowledge Base

Browsing the Knowledge Base. The knowledge base collects an amount of knowledge objects (which consist of the related data elements) and category objects (which consist of data about the subjects of the categories). One of browsing the knowledge base is to render the categories as trees and category objects and knowledge objects as tree nodes. By using the command `browseBy[`*objectID*`]` where *objectID* is the *objectID* of a category object, the system will retrieve the *objectIDs* of its members based on the categoryStructure table (see Fig. 5) and perform the same process on these members recursively until they are all knowledge objects. According to these *objectIDs*, the tree nodes will be generated by fetching values of the attribute *name* stored in the knowledgeName table (see Fig. 2) and the categoryObject table (see Fig. 4).

The other view is to browse the data within knowledge objects and category objects. Given an *objectID*, the system automatically generate corresponding XML documents by assembling the related data with this *objectID* through the structure within the objects (see Figs. 1, 2, and 4) and render them in readable styles via SAXON XSLT processor [9] and JDesktop Integration Components (JDIC [6]), which provides Java applications with access to functionalities and facilities furnished by the native desktop (see Fig. 8).

Querying the Knowledge Base. Currently, the system provides basic query services through keywords and relations for users to input search commands and to view the results. The queries through relations work at the level of knowledge objects and category objects. This means that the queries need be described by using the *objectIDs* of the objects but not simple natural texts. We explain the commands for queries as follows.

- `keyWords[`$word_1, \ldots, word_n$`]` returns the set of knowledge objects and category objects with keywords $word_1$ and ... and $word_n$ according to the keyWords table (see Fig. 2).
- `relation[*,`*objectID*`,`*relationType*`]` returns the set of knowledge objects such that their subsequences are the knowledge object identified by *objectID* with the relation of *relationType*; `relation[`*objectID*`,*,`*relationType*`]` returns the set of knowledge objects such that their precursors are the knowledge objects identified by *objectID* with the relation of *relationType*. These two query commands work by using the KOstructure table (see Fig. 5).

The provided commands support fetching the knowledge objects related with a given knowledge object and are helpful for users to author geometric documents. For example, if one wants to include "Simson's theorem" into a document, he/she can query the knowledge base through keywords to see whether others have

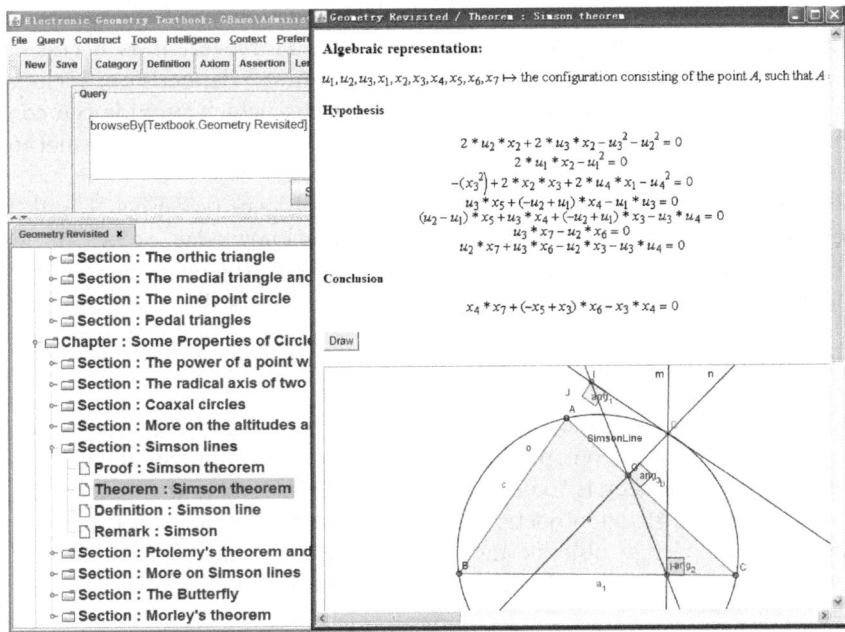

Fig. 8. Browsing the content of "Simson's theorem" object

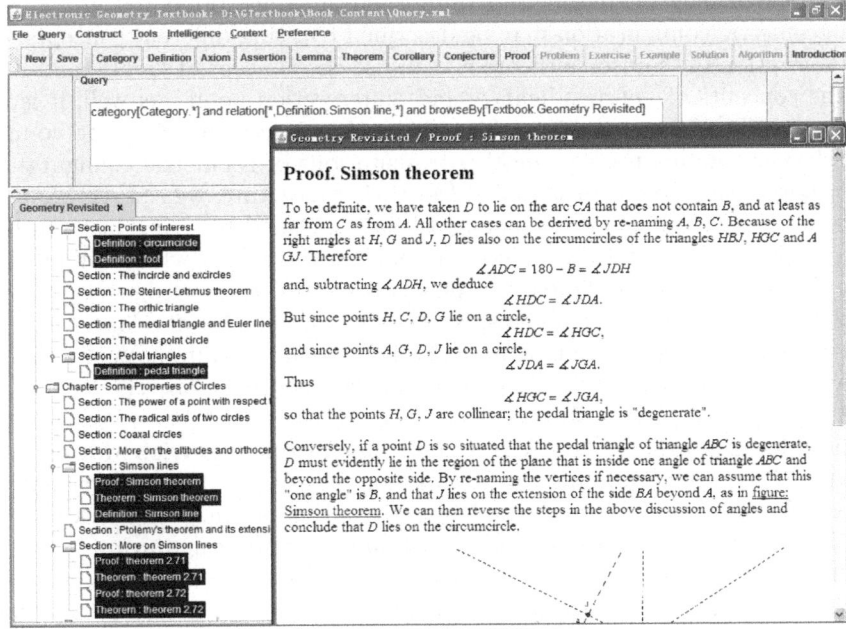

Fig. 9. Querying the knowledge base

constructed the theorem; if so, he/she can browse the data of the theorem and modify the data to satisfy his/her requirement. Moreover, it is necessary to include the knowledge objects that "Simson's theorem" depends on, such as the definitions of "foot", "collinear", and "circumcircle" which provide the context of "Simson's theorem". To sort the results of retrieval, we adopt the method of viewing them in the structure of a given category (see Fig. 9).

According to the structure within knowledge objects (see Figs. 1 and 2), it is possible to retrieve the needed data within a knowledge object by giving the *objectID* of the knowledge object, as described in the example in Sect. 2.3. However, the specification of this functionality or API is still under development.

5   Conclusion and Future Work

In this paper, we have discussed our ideas on the design of a geometric knowledge base and presented our preliminary implementation of the knowledge base system. Our key strategy is to encapsulate certain interrelated geometric knowledge data into knowledge objects. We have focused our study on the hierarchic structure of knowledge objects and its formalization and representation. The application tools to realize the automated activities proposed in Sect. 2.1 are being developed.

So far the functionality of query is limited as we have to know the *objectID* of the concerned knowledge object in the knowledge base first. The challenge is to search and sort the knowledge objects that match a given geometric configuration which may be specified in the formal language presented in Sect. 2.2. As a geometric configuration may be described by using different concepts, e.g., the description of "Simson's theorem" can use the concept of "foot", and it can use the concepts of "perpendicular" and "intersection point" as well, it is necessary to define a canonical form for each geometric configuration and compute the relevance among the canonical representations of geometric configurations. According to the Type structure and Derivation structure, we can see what concepts are basic. The geometric configurations specified by using just the basic concepts may approach to the canonical form for retrieving, which will be investigated at the next stage. Moreover, the structure of the knowledge base, i.e., the relations among knowledge objects, is annotated by human developers or users. It is valuable to investigate how to acquire the relations automatically through analyzing and mining the data contained in the knowledge objects.

Acknowledgments. The authors wish to thank Professor Arjeh M. Cohen and Dr. Hans Cuypers for offering the first author the opportunity to visit and work in their DAM group at Technische Universiteit Eindhoven. This work has benefited considerably from the friendly environment of working during his stay in Eindhoven. The authors also wish to thank the referees for their insightful comments and suggestions which have helped bring this paper to the present form. The research has been supported by the Chinese National Key Basic Research (973) Project 2005CB321901/2 and the SKLSDE Open Fund BUAA-SKLSDE-09KF-01.

References

1. Buswell, S., Caprotti, O., Carlisle, D., Dewar, M., Gaëtano, M., Kohlhase, M.: The OpenMath Standard, Version 2.0. Tech. Rep., The OpenMath Society (2004), http://www.openmath.org/
2. Chen, X., Wang, D.: Towards an Electronic Geometry Textbook. In: Botana, F., Recio, T. (eds.) ADG 2006. LNCS (LNAI), vol. 4869, pp. 1–23. Springer, Heidelberg (2007)
3. GeoGebra Home, http://www.geogebra.org/cms/
4. Gräbe, H.-G.: The SymbolicData GEO Records – A Public Repository of Geometry Theorem Proof Schemes. In: Winkler, F. (ed.) ADG 2002. LNCS (LNAI), vol. 2930, pp. 67–86. Springer, Heidelberg (2004)
5. Intergeo Home, http://i2geo.net/xwiki/bin/view/Main/
6. JDIC Home, https://jdic.dev.java.net/
7. MathDox Formula Editor, http://www.mathdox.org/formulaeditor/
8. Quaresma, P., Janičić, P.: GeoThms — Geometry Framework. Tech. Rep. CISUC TR 2006/002, Centre for Informatics and Systems, University of Coimbra (2006)
9. SAXON XSLT Processor Home, http://saxon.sourceforge.net/
10. Wang, D.: GEOTHER 1.1: Handling and Proving Geometric Theorems Automatically. In: Winkler, F. (ed.) ADG 2002. LNCS (LNAI), vol. 2930, pp. 194–215. Springer, Heidelberg (2004)
11. Wang, D.: Formalization and Specification of Geometric Knowledge Objects. In: Hu, Z., Zhang, J. (eds.) Proceedings of the Sixth Asian Workshop on Foundations of Software (AWFS 2009), Tokyo, Japan, pp. 86–98 (2009)

Proof Certificates for Algebra
and Their Application to
Automatic Geometry Theorem Proving

Benjamin Grégoire, Loïc Pottier, and Laurent Théry

Marelle Project, INRIA Sophia Antipolis

Abstract. Integrating decision procedures in proof assistants in a safe way is a major challenge. In this paper, we describe how, starting from Hilbert's Nullstellensatz theorem, we combine a modified version of Buchberger's algorithm and some reflexive techniques to get an effective procedure that automatically produces formal proofs of theorems in geometry. The method is implemented in the Coq system but, since our specialised version of Buchberger's algorithm outputs explicit proof certificates, it could be easily adapted to other proof assistants.

Keywords: decision procedure, Nullstellensatz, geometry theorem proving, proof assistant.

1 Introduction

Integrating decision procedures in proof assistants in a safe way is a major challenge. Many well-known and widely-used decision procedures exist but making them available in the context of a proof assistant is not so trivial: one has to certify the result of the procedure. This may explain, for example, why very few theorems of geometry have been formalised yet in the list compiled by Freek Wiedijk [28].

The integration can be made in several ways. Everything can be done inside the proof assistant. For example, one can write the procedure as a tactic, then, every time the tactic is called, a proof is built for the particular instance. Alternatively, if the system offers a programming language, one can write the entire procedure as a program inside the system and use standard program verification techniques to derive its correctness once and for all. In both cases, such an internal integration is usually very time consuming specially if the procedure is rather complex.

Another way to go is to use external programs. For example, one can take an existing implementation of the procedure and modifies it to output execution traces. Then, the proof assistant only needs to follow the information in the trace to build its own proof. An alternative is to use certificates instead of traces. Certificates do not contain all the execution path but just enough information to make it easy for the proof assistant to build a proof. Examples of such certificates are prime certificates that we have used [13] to get the primality of

T. Sturm and C. Zengler (Eds.): ADG 2008, LNAI 6301, pp. 42–59, 2011.
© Springer-Verlag Berlin Heidelberg 2011

large prime numbers, or algebraic certificates [17]. Verifying certificates is usually more difficult than checking traces but still an order of magnitude simpler than building internally the entire procedure. Verifying the certificates can be done either by tactics or by a verified program.

In this paper, we advocate the combination of an external program that generates certificates and a verified program that checks certificates for the particular case of deciding problems of the following kind:

$$\forall X_1, \ldots, X_n \in R,$$
$$P_1(X_1, \ldots, X_n) = 0 \wedge \ldots \wedge P_s(X_1, \ldots, X_n) = 0$$
$$\Rightarrow P(X_1, \ldots, X_n) = 0$$

where R is a commutative ring without zero divisor and $P, P_1, \ldots, P_s$ are polynomials. This has been intensively studied during the 80s, specially by the computer algebra community. The main decision procedure used to solve these problems is Buchberger's algorithm [2]. A lot of efforts have been spent in improving this algorithm and trying to apply it to various domains of theorem proving. Without doubt, one of the most successful area of application is automatic geometry theorem proving (see for example [5,20,21,26,29,22] for a survey). In the context of proof assistants, this has also recently drawn some attention. John Harrison [16] uses a basic implementation of Buchberger's algorithm to produce a certificate for elementary arithmetic (e.g. proving the Chinese remainder theorem) for the HoLLight system [14]. Also, Amine Chaieb and Makarius Wenzel [3] use it in the context of the Isabelle system [24]. One of the authors [25] has also connected Coq with the state-of-the-art implementation of Gröbner bases algorithms.

In this paper, we explain how we manage to solve theorems of geometry proved by Wu's method: Desargues, Pascal, and 20 other theorems. Getting these theorems inside a proof assistant is a real challenge: the proof certificates we obtain for these theorems are quite large. Three key ingredients were essential:

- our implementation of Buchberger's algorithm that generates certificates does not try to compute the whole Gröbner basis. It just does enough work to solve the specific problem. In practice, this reduces drastically the time that is needed to generate certificates. Our approach is very pragmatic and lacks a more precise insight of what is exactly gained by progressively reducing the polynomials.
- our certificates are not just composed of a single polynomial identity but consist in *straight-line programs*. Such programs are composed of assignments only. No branching or loops are allowed. Straight-line programs are important tools used in algebraic complexity to prove optimal bounds [9]. Here, they appear to be also of great practical use.
- reflexive methods and the power of the reduction machine of the Coq system are used to verify these certificates efficiently.

A tactic in Coq is a piece of program that allows to prove automatically some specific kind of logical statements. It produces a proof term for this statement. The proof is then verified by Coq. If it is correct, the statement is validated as a theorem. In this work, we have implemented and tested a new tactic for proving ideal membership. Its input is a statement with polynomial expressions as hypotheses and conclusion. Its output is a proof term that is automatically generated from a certificate (a list of polynomials lists). Another tactic has been developed for geometry. It reduces geometrical statements into their polynomial form. We can then compose the two tactics to prove geometrical statements.

The paper is organised as follows. In Section 2, we recall what the Nullstellensatz theorem and Gröbner bases are. In Section 3, we present our modified version of Buchberger's algorithm that generates certificates based on straight-line programs. In Section 4, we recall what reflexive methods in provers like CoQ are and explain how they have been used in our context to produce short proof terms. Finally, in Section 5, we illustrate an application of our decision procedure to prove automatically some theorems of geometry taken from [5] and [6].

2   Nullstellensatz Theorem and Gröbner Basis

We seek to prove implications of the following form:

$$\forall X_1, \ldots, X_n \in R,$$
$$P_1(X_1, \ldots, X_n) = 0 \land \ldots \land P_s(X_1, \ldots, X_n) = 0$$
$$\Rightarrow P(X_1, \ldots, X_n) = 0$$

where R is a commutative ring without zero divisor and $P_1, \ldots, P_s$ are polynomials. As a matter of fact, the general problem we want to solve is quantifier elimination in general rings and fields, but the general problem reduces easily to the problem we address here (see for example [23]). Hilbert's Nullstellensatz theorem shows how to reduce proofs of equalities on polynomials to algebraic computations (see for example [7] for the notions introduced in this section).

It is easy to see that if a polynomial P in $R[X_1, \ldots, X_n]$ verifies $cP^r = \sum_{i=1}^{s} Q_i P_i$, with $c \in R$, $c \neq 0$, r a positive integer, and the Q_is in $R[X_1, \ldots, X_n]$, then P is zero whenever polynomials $P_1, \ldots, P_s$ are zero. The converse is also true when R is an algebraic closed field: the method is complete. So, proving our initial problem reduces into finding $Q_1, \ldots, Q_s$, c and r such that $cP^r = \sum_i Q_i P_i$. In this case, we call $(c, r, Q_1, \ldots, Q_s)$ the "certificate" of the statement we want to prove since it is straightforward to obtain a proof from this certificate.

In this work, we concentrate on the special case where $r = 1$, i.e. the problem of finding $Q_1, \ldots, Q_s$ and c such that $cP = \sum_i Q_i P_i$. The cases $r > 1$ can be tested by enumeration, or by using extra variables, as explained in [25] for example. In practice, almost all problems are solved with $r = 1$.

2.1   Division of Polynomials

An *ideal* $\mathcal{I}$ of a ring is an additive subgroup of the ring such that $a \times x \in \mathcal{I}$ whenever $a \in \mathcal{I}$. The ideal *generated* by a family of polynomials is the set of all linear combinations of these polynomials (with polynomial coefficients). A *Gröbner basis* of an ideal is a set of polynomials of the ideal such that their head monomials (relative to a chosen order on monomials, e.g. lexicographic order, or degree order) generate the ideal of head monomials of all polynomials in the ideal. The main property of a Gröbner basis is that it provides a test for the membership to the ideal: a polynomial is in the ideal if and only if its euclidean *division* by the polynomials of the basis gives a zero remainder.

The division process is a generalisation of the division of polynomials in one variable: to divide a polynomial P by a polynomial $aX^\alpha - Q$ we write $P = aX^\alpha S + T$ where T contains no monomial that is multiple of X^α. Then we change P into $QS + T$ and repeat the process. When we reach a polynomial that is not divisible by $aX^\alpha - Q$, this is the remainder of the division. For example, suppose that we use the degree order on monomials. The head monomial of $x^2 - z$ is then x^2. In order to divide $x^4y + x^2 - 1$ by the polynomial $x^2 - z$, we rewrite x^2 into z everywhere. This leads to the polynomial $z^2y + z - 1$, which is not divisible by $x^2 - z$: it is the remainder of this division process. In order to divide a polynomial by a family of polynomials, we repeat this process with each polynomial of the family.

2.2   Gröbner Bases

In general, the remainder of a division of a polynomial by a family of polynomials depends on the order in which we use the polynomials of the family. With a Gröbner basis, this remainder is unique: this is a characteristic property of Gröbner bases. For example, dividing $x^2y^2 - y^4$ by the family $\{x^2 + 1, xy - 1\}$ gives the remainder $-y^2 - y^4$ if we divide by $x^2 + 1$ but gives $1 - y^4$ if we divide by $xy - 1$. Both remainders are irreducible: so the family is not a Gröbner basis. We can prove that the family $\{x^2 + 1, xy - 1, x + y, y^2 + 1\}$ is a Gröbner basis of the ideal generated by $\{x^2 + 1, xy - 1\}$. Any division of $x^2y^2 - y^4$ by this family will give the remainder 0. With a simple division algorithm, we can conclude that the polynomial $x^2y^2 - y^4$ is in the ideal generated by $\{x^2 + 1, xy - 1\}$.

Consider now polynomials of the ideal with a degree in y strictly lower than 2 such as $x^3 - y$. Obviously, dividing them by only the three first polynomials $\{x^2 + 1, xy - 1, x + y\}$ of the basis is sufficient to give the remainder 0. This shows that on some particular cases it is not necessary to have a complete Gröbner basis in order to conclude. Also, half of the time for computing a Gröbner basis is usually spent in verifications that do not produce any new elements for that basis. In practice, the strategy that consists in checking the membership each time a new polynomial is added to the family gives an effective speed-up.

3   Buchberger's Algorithm and Certificates

The main method for computing Gröbner bases is Buchberger's algorithm. It consists in completing the initial family of polynomials by new polynomials in the same ideal built from the so-called *S-polynomials*. For a pair of polynomials (P, Q), its associated S-polynomial is the polynomial $t_1 P - t_2 Q$ where the terms t_1 and t_2 are chosen in such a way that the head monomials of $t_1 P$ and $t_2 Q$ are identical, so they cancel out in the subtraction. The algorithm starts with the initial family and adds all the non-zero remainders of the S-polynomials for all pairs of elements of the family. If no new element has been added to the family, the algorithm terminates otherwise it repeats the same completion process with the new family. Dixon's lemma ensures that this iterative process eventually terminates. The resulting family is then a Gröbner basis.

We modify this algorithm in a simple way: each time a non-zero remainder of a S-polynomial is added to the family, we divide the polynomial P by it, and replace P by its remainder. If this remainder is zero, the algorithm terminates. Remembering all the divisions that has been done from the beginning gives a way of writing P as a linear combination of the original polynomials. Let us take a concrete example. Suppose we want to show that $P_1 = 0 \land P_2 = 0 \Rightarrow P = 0$ with $P_1 = x^2 + 1$, $P_2 = xy - 1$ and $P = x^3 - y$. Our initial family is then $\{P_1, P_2\}$ and we want to find a certificate c, Q_1 and Q_2 such that $cP = Q_1 P_1 + Q_2 P_2$. We first try to divide P by the family $\{P_1, P_2\}$ starting from left to right. P_1 divides P and the remainder is $R_1 = -x - y = P - xP_1$. R_1 is non-zero and irreducible by $\{P_1, P_2\}$, so we can start the completion. There is only one S-polynomial for the family $\{P_1, P_2\}$, which is $P_3 = x + y = yP_1 - xP_2$. It is irreducible by $\{P_1, P_2\}$, so we add $P_3 = x + y$ to the family $\{P_1, P_2\}$. We then try to divide R_1 by P_3, this gives $0 = R_1 + P_3$ so we can stop the completion. We have

$$0 = R_1 + P_3 = (P - xP_1) + (yP_1 - xP_2)$$

singling P out gives

$$P = (x - y)P_1 + xP_2$$

so the certificate is $c = 1$, $Q_1 = x - y$ and $Q_2 = x$. Note that in order to get the certificate, we have not computed the complete basis.

For our certificates, we are not going to express P as a combination of the initial family as in the previous example. A more effective way of presenting the certificate is to be a bit closer to the computation that is actually performed by the algorithm. The certificate is then composed of two parts. The first part, CR, is a list of polynomials. It gives the coefficients to express P as a combination of the initial family plus the extra polynomials that the computation of the partial Gröbner basis has added to the initial family in order to reduce P to 0. The second part, C, is a list of polynomials lists. Each subsist corresponds to one extra polynomial and explains why it belongs to the ideal generated by the initial family plus the polynomials that are added before. More formally, with an initial family $\{P_1, \ldots, P_s\}$ this gives:

$$CR = [c_1, \ldots, c_{s+p}]$$
$$C = [[a_{1\ s+1}, \ldots, a_{s\ s+1}],$$
$$\ldots$$
$$[a_{1\ s+p}, \ldots, a_{s\ s+p}, \ldots, a_{s+p-1\ s+p}]]$$

where

$$\forall i \in [1; p], P_{s+i} = a_{1\ s+i}P_1 + \ldots + a_{s+i-1\ s+i}P_{s+i-1}$$

and

$$P = -(c_1 P_1 + \ldots + c_{s+p} P_{s+p})$$

For simplicity here, we have assumed that R is a field (hence $c = 1$) but we can easily extend the certificate format to the case where divisions are pseudo-divisions. As each new polynomial in C is defined with respect to the previous ones, C has a structure that is very similar to straight-line programs. Applied to our example, we get the following "program":

$$P_1 := x^2 + 1;$$
$$P_2 := xy - 1;$$
$$P_3 := yP_1 + (-x)P_2;$$
$$P := -((-x)P_1 + 0P_2 + 1P_3);$$

so the certificate is $CR = [-x, 0, 1]$ and $C = [[y, -x]]$. The main advantage of straight-line programs is that they allow the sharing of computations. This can change the exponential computing time into linear one (see [9] for an example of how straight-line programs can be used to find complexity bound).

In general, a straight-line program is an imperative program without loops, i.e. a sequence of assignments of expressions to variables, each expression depending on previously assigned variables:

$$x_1 := f_1();$$
$$x_2 := f_2(x_1);$$
$$x_3 := f_3(x_1, x_2);$$
$$\ldots$$
$$x_n := f_n(x_1, \ldots, x_{n-1});$$

where $f_1, f_2, \ldots, f_n$ are parametric procedures. In computer algebra, they are usually rational fractions, here just polynomials. A straight-line program can be viewed as a directed acyclic graph, i.e. a tree with shared sub-trees.

To illustrate how this change of complexity may occur in our particular context, let us consider the following contrived example. Let f_n be the nth Fibonacci number: $f_0 = 0, f_1 = 1, f_{n+2} = f_{n+1} + f_n$ and suppose that we want to prove that $X^{f_{n+1}} - 1 = 0 \wedge X^{f_n} - 1 = 0 \Rightarrow X - 1 = 0$. Computing the Gröbner basis of $\{X^{f_{n+1}} - 1, X^{f_n} - 1\}$ mimics Euclid's algorithm for gcd and the decomposition is

$$X - 1 = P_{n-2}(X^{f_{n+1}} - 1) + P_{n-1}(X^{f_n} - 1) \tag{1}$$

where the polynomials P_n are defined by $P_0 = 0, P_1 = 1, P_n = -X^{f_n} P_{n-1} + P_{n-2}$. The certificate with straight-line programs is

$$CR = [0, \ldots, 0, 1]$$
$$C = [[1, -X^{f_{n-1}}],$$
$$[0, 1, -X^{f_{n-2}}], \tag{2}$$
$$\ldots$$
$$[0, \ldots, 0, 1, -X^{f_2}]]$$

```
inideal(P,F){
   (* F = [P1,...,Pn] *)
   R := P; C := [];CR := LR; R := R1;
   SP:= Spolynomials(F,F);
   let (R1,LR) = divide(R,F) in
   while R <> 0 do (* stop if P divides to 0 by F *)
     if SP = [] then Fail
        (* the Gröbner basis is computed without reducing P to 0 *)
     else let (S,LS)::SP1 = SP in
           SP := SP1;
           let (D,LD) = divide(S,F) in
           if D <> 0 (* add a new polynomial to F *)
              then F := F + [D];   (* + denotes concatenation of lists *)
                   SP := SP + Spolynomials(F,[D]);
                   C := C + [merge(LD,LS)];
                   let (R1,LR) = divide(R,F) in (* reduce R by F *)
                   CR := merge(CR,LR);
                   R := R1;
   done;
   return(CR,C);
}
```

Fig. 1. Pseudo-code of the modified Buchberger's algorithm with certificate

Suppose that in order to check a polynomial equality, one first applies distributivity and then collects equal monomials. Let us compare the verification of the two certificates (1) and (2) in term of operations on monomials. In the first one, P_n has degree $f_{n+2} - 2$ and has f_n monomials, with coefficients 1 (n odd) or -1 (n even), then the verification of this equation requires $2f_n$ multiplications, and $3f_n$ additions. In the second one, each intermediate polynomial has form $X^{f_k} - 1$, then the verification only requires $2n - 4$ multiplications and $4n - 8$ additions. As $f_n \sim ((1 + \sqrt{5})/2)^n/\sqrt{5}$, there is an exponential factor between the two.

We end this section by giving in Figure 1 the pseudo-code of the function inideal that generates our certificates. The function divide returns the remainder together with the list of quotients of the division. The function Spolynomials computes the S-polynomials of two families. For each of these S-polynomials, it also returns the monomials and polynomials used to compute them. The function merge adds terms of two lists of same rank, completing by zeros if needed. Our program is composed of 3500 lines of OCAML and 500 lines of COQ. It includes polynomial arithmetic for sparse polynomials and recursive polynomials, rational fractions, sub-resultant, gcd computation, and unbounded integer arithmetic. The OCAML code could easily be used as a standalone prover provided one adds a minimal parser/printer for polynomials.

4   Reflexive Method to Verify Large Certificates in Proof Assistant

In COQ system, each deduction step appears explicitly in the final proof term. Thus, each application of lemma (and in particular each rewriting step) is stored in the proof. This clearly prohibits the use of rewriting tactics to verify our certificates: proof terms would be too large. Fortunately, the COQ system integrates a programming language on which we can reason. For programs written in this language, symbolic evaluation is also possible via the reduction mechanism. More importantly, this reduction mechanism is integrated inside the logic: two terms are considered equal if their normal forms, i.e. the terms after evaluation, are structurally equal. So, reductions do not appear in proof terms. The reflexive method introduced by Allen et al. [1] takes advantage of this reduction mechanism in order to reduce drastically the size of proof terms. Note that if the reduction mechanism is particularly efficient, using reflexive methods can also reduce the time required to verify the proof. The reflexive method relies on the following remark:

- Let $P : A \rightarrow$ Prop be a predicate over a set A.
- Assume we are able to write in the system a program c such that the following properties holds

$$\text{c\_spec} :\ \forall a,\ c\ a = \text{true} \rightarrow P\ a$$

In other words, for all value a, if the evaluation of the program c on a returns true then P is satisfied for a. This means that c is a semi decision procedure for the properties P and c_spec is the lemma which expresses that the semi decision procedure is correct.

Now, assume that we have to prove $P\ a'$ for a specific a' and that, for this particular a', the system is able to reduce $c\ a'$ into true. In order to prove $P\ a'$, we can apply the lemma c_spec to a', so we are left with $c\ a' = \text{true}$ to prove. Since $c\ a'$ reduces to true, this proposition $c\ a' = \text{true}$ is identical for the prover to the proposition $\text{true} = \text{true}$ which can be proved by the reflexivity of equality.

The COQ system is based on the Curry-Howard isomorphism. This means that proofs are represented by programs, propositions by types and valid proofs are well-typed programs. For example, our proof of $P\ a'$ is the program c_spec a' (refleq true). The typing derivation of the proof is:

$$\cfrac{\cfrac{}{\Gamma \vdash \text{c\_spec } a' : c\ a' = \text{true} \rightarrow P\ a'}\ \vdots \qquad \cfrac{\cfrac{\Gamma \vdash \text{refleq true} : \text{true} = \text{true} \qquad \text{true} = \text{true} \equiv c\ a' = \text{true}}{\Gamma \vdash \text{refleq true} : c\ a' = \text{true}}\text{[Conv]}}{}}{\Gamma \vdash \text{c\_spec } a'\ (\text{refleq true}) : P\ a'}$$

Here the key point is the use of the CONV typing rule:

$$\cfrac{\Gamma \vdash t : T \qquad T \equiv U}{\Gamma \vdash t : U}\text{[Conv]}$$

which allows to view a program t of type T as a program of type U if T and U are equal modulo reduction (convertible). All the reduction steps that are necessary to check the convertibility of $\text{true} = \text{true}$ and $c\ a' = \text{true}$ do not appear in the proof term. Naturally, if the reduction steps are not in the proof term, they are going to be performed during the checking phase. So the time needed to check a proof that uses the reflexive method will not only crucially depend on the efficiency of the reduction mechanism implemented by the prover but also on the efficiency of the semi-decision procedure. The compiled reduction mechanism of CoQ has a very efficient strategy to reduce programs [11]. This makes the reflexive method very attractive in CoQ.

Now that we have introduced the reflexive method, let us explain how it can be used to define a checker for the certificates defined in Section 3. For polynomial operations, we use the existing polynomial library of CoQ [12]. This library has been developed for the reflexive ring tactic that proves equalities over an arbitrary ring structure. It defines two data-types:

- $\mathcal{E}$ represents the type of polynomial expressions, i.e. the free algebra;
- $\mathcal{P}$ represents the type of polynomials in Horner normal form.

Basic operations like addition or multiplication are defined on the type $\mathcal{P}$, thus it is easy to define a normalisation algorithm norm from $\mathcal{E}$ to $\mathcal{P}$ by structural recursion. Correctness of the basic operations is provided using an interpretation function $[\![\ ]\!]_\rho$ from $\mathcal{P}$ to an arbitrary ring R, where ρ is the valuation function that binds polynomial indeterminates to value in R. It is proved that each operator is correct with respect to the interpretation function. For example the specification of the addition is given by:

$$\forall \rho\ P_1\ P_2,\ [\![P_1 +_{\mathcal{P}} P_2]\!]_\rho = [\![P_1]\!]_\rho +_R [\![P_2]\!]_\rho$$

In a similar way, the correctness of the normalisation is defined using an interpretation function $[\![\ ]\!]_\rho^{\mathcal{E}}$ on polynomial expressions:

$$\forall \rho\ E,\ [\![E]\!]_\rho^{\mathcal{E}} = [\![\text{norm } E]\!]_\rho$$

Thus, to prove that two ring expressions r_1 and r_2 in R are equal, it is sufficient to find two polynomial expressions E_1, E_2 and a valuation function ρ such that $[\![E_i]\!]_\rho^{\mathcal{E}}$ reduce to r_i. If the normalisation of E_1 and E_2 leads to the same Horner normal form then r_1 and r_2 are equal. This strategy is not necessarily the best one. The normalisation is defined by a naive structural recursion: in order to normalise $(X + Y)^{100} - (X + Y)^{100}$ the function first normalises twice the sub-term $(X + Y)^{100}$ and then performs the subtraction. This is clearly not optimal.

Implementing our checker on top of that library is straightforward. We first define a function mult_1 that normalises each line of the certificate. In the syntax of CoQ this looks like

```
Function mult_1 (Lₑ L: list 𝒫) : 𝒫 :=
  match Lₑ, L with
  | e::L'ₑ, p::L' ⇒ e *_𝒫 p +_𝒫 mult_1 L'ₑ L'
```

```
| _, _ ⇒ 0_P
end.
```

Then a second function `compute_list` collects all the normalised polynomials that correspond to the lines of the certificates

```
Function compute_list (LL_e: list (list P)) (L:list P): list P :=
  match LL_e with
  | L_e::LL_e ⇒ compute_list LL_e ((mult_1 L_e L)::L)
  | _ ⇒ L
  end.
```

Finally the checking function `check` tests the equality of the two normal forms:

```
Function check (L_e:list E) (p:E) (certif: list (list P) * list P) :=
  let (LL_e, L'_e) := certif in
  let L := map norm L_e in
  norm p =?=_P mult_1 L'_e (compute_list LL_e L).
```

Note that all the functions we have defined for our checker are tail-recursive. In order to prove the correctness of the checker, we first define the property for a list to be composed of only zero polynomials:

```
Definition Allzero ρ (L: list P) := ∀P ∈ L, [[P]]_ρ = 0.
Definition Allzero_E ρ (L_e: list E) := ∀P ∈ L_e, [[P]]^E_ρ = 0.
```

We then show that the two functions `mult_1` and `compute_list` behave well with list of zero polynomials:

```
Lemma mult_1_spec: ∀ρ L_e L, Allzero ρ L → [[mult_1 L_e L]]_ρ = 0.
Lemma compute_list_spec:
  ∀ρ LL_e L, Allzero ρ L → Allzero ρ (compute_list LL_e L).
```

Finally, we can derive the correctness of our checker

```
Lemma check_correct:
  ∀ρ L p certif, check L p certif = true → Allzero_E ρ L → [[p]]^E_ρ = 0.
```

Defining the checker and proving its correctness is straightforward. This is exactly what we wanted: the integration of a decision procedure from the prover side should be as seamless as possible.

5 Geometry Theorem Proving

In his book [5], Shang-Ching Chou proves 512 theorems of geometry mechanically. In this section, we show how we have been capable to prove some of the most difficult ones in CoQ with our certificates. We also compare our results with other systems (HOL Light [14] and Macaulay2 [10]).

In order to turn geometry into algebra, points are represented by their coordinates, geometric predicates by polynomials based on determinants, scalar products and algebraic relations between trigonometric functions. For example, we define collinearity in CoQ by:

```
Definition collinear (A B C:point):=
  (X A - X B) * (Y C - Y B) - (Y A - Y B) * (X C - X B) = 0.
```

and the fact two lines defined by two pairs of points are parallel:

```
Definition parallel (A B C D:point):=
  (X A - X B) * (Y C - Y D) = (Y A - Y B) * (X C - X D).
```

Figure 2 gives a summary of some of our experiments. The machine used for these benchmarks is a Linux PC with dual Intel Xeon 3.2Ghz processors with 33Gb of memory. Columns contain respectively:

1. The name of the theorem and in parenthesis the page in Chou's book where it is stated (when it exists);
2. The time in seconds for computing the certificate,
3. The time in seconds for verifying the certificate,
4. The size in number of characters of c and the Q_i when expanding the certificate into $cP = \sum_i Q_i P_i$.
5. The size in number of characters of the certificate,
6. The size of the certificate as a proof term (number of nodes),
7. The size of the certificate as a optimized straight-line program: every subterm of the proof term is shared (with *let* operator).

A more detailed presentation of the examples is available at

$$\texttt{http://www-sop.inria.fr/marelle/CertiGeo}$$

These results deserve some comments:

- The number of variables of the Gröbner bases computation is about 20 (the number of coordinates of the points) and the degree of the input polynomials is generally 2 (which is the general case: each ideal can be generated by polynomials of total degree less than two, provided we add extra variables).
- The time for computing a Gröbner base is very sensitive to the variable order. In general, a good choice is to have the variables of base points greater than the variables of constructed points, and to use a reverse-lexicographic term order. But this is not always the case. Only when computing with this naive order was prohibitive, we did try to find an better order. The names of such theorems are marked with a star in Figure 2.
- Certificates with straight-line programs are generally better than raw polynomials. For examples like Ceva's theorem, this makes a significant difference.

Theorems in geometry are not true in general, there are some non-degeneracy conditions: some particular points must not be collinear, some lines not parallel, and so on. From the algebra point of view, this means that we can only prove:

$$CP = \sum_i Q_i P_i$$

where C is a polynomial in some variables, which are parameters of the theorem. From a logical point of view, this means that the conclusion of the theorem

Theorem	Time (seconds)		Size (characters)		Size (nodes)	
	Computing	Verifying	Polynomials	Certificate	Term	SLP
Ceva (264)	181	2.5	538644	477414	266233	76669
Desargues (*)(269)	0.3	0.01	6359	4551	24527	4311
Feuerbach (199)	0.8	0.4	52569	16999	15585	5497
Pappus (*)(100)	1.3	0.2	2721	1934	29945	8031
Pascal_circle (*)	397	12	732982	864509	754290	183505
Pascal_circle2 (20)	91	2.6	10603	15128	312626	66154
Ptolemy (*)	1.1	0.5	1549	1556	26210	9129
Ptolemy_theo95 (142)	200	2.4	571931	571931	344278	73257
Pythagora	0.000	0.009	7	7	4	4
Simson (240)	0.3	0.2	1541	1238	15680	4919
Thales	0.03	0.1	5422	5169	3146	1323
bisectors	0.002	0.04	165	165	105	69
butterfly (119)	0.1	0.2	12116	11125	13661	3980
Euler circle	0.06	0.5	5532	2936	2795	1146
chords	0.002	0.04	639	642	568	282
altitudes	1.1	0.3	4947	5386	5295	1801
isosceles	0.001	0.01	10	10	3	3
medians	0.005	0.06	2910	2717	2284	1064
bisections	0.005	0.06	2577	2145	1911	831
Minh	0.07	0.1	3367	3616	3987	1881
SegmentsofChords	0.1	0.09	10375	9839	7476	2491
threepoints	0.11	0.13	2890	2587	2796	1105
fib(16)	0.003	0.8	15786	393	416	137
fib(17)	0.004	1.3	26059	423	465	151
fib(18)	0.004	2.4	40864	464	517	166
fib(22)	0.008	68	307720	630	753	225

Fig. 2. Times and sizes of some selected theorems

becomes a disjunction of the original conclusion and the degenerate cases. A work-around is to work with coefficients that are rational fractions in some variables $u_1, \ldots u_r$, called *parameters*. Polynomials are then not in $R[X_1, \ldots, X_n]$ anymore but in $R(u_1, \ldots, u_r)[X_1, \ldots, X_n]$. Let us take for example Desargues' theorem. It states that given two triangles (A, B, C) and $(A1, B1, C1)$ and a point S such that S, A, $A1$ are collinear, A, B, $B1$ are collinear, and S, C, $C1$ are collinear, we can deduce that the intersection R of (A, B) and $(A1, B1)$, the intersection Q of (A, C) and $(A1, C1)$, and the intersection P of (B, C) and $(B1, C1)$ are also collinear. All points are in the affine plane. Its statement and proof in CoQ are:

```
Lemma Desargues: forall A B C A1 B1 C1 P Q R S:point,
  X S = 0 -> Y S = 0 -> Y A = 0 ->
  collinear A S A1 -> collinear B S B1 -> collinear C S C1 ->
  collinear B1 C1 P -> collinear B C P ->
  collinear A1 C1 Q -> collinear A C Q ->
  collinear A1 B1 R -> collinear A B R -> collinear P Q R
```

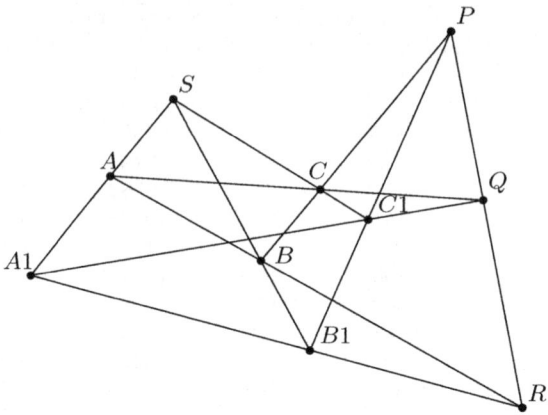

Fig. 3. Desargues'theorem

```
 ∨ X A = X B ∨ X A = X C ∨ X B = X C ∨ X A = 0
 ∨ collinear S B C
 ∨ parallel A C A1 C1 ∨ parallel A B A1 B1.
Proof.
geo_begin.
tzRpv 0%Z (X A::X B::Y B::X C::Y C::X A1::Y B1::Y C1::nil)
 (X B1::X C1::Y P::X P::Y Q::X Q::Y R::X R
 ::Y C1::Y B1::X A1::Y A1::Y C::X C::Y B::X B::nil).
Qed.
```

The theorem is proved by two tactics. The first one, geo_begin, transforms the statement in an algebraic one (disjunctions in conclusion become products, negations in hypothesis like $p \neq 0$ becomes $t * p = 1$ where t is a new variable and so on). The second one, tdzRpv, takes explicitly as arguments the parameters and the variables with which the certificate generator has to be called and checks back the resulting certificate. Without the extra conditions X A = X B ∨...∨ parallel A B A1 B1, the theorem is not true. In order to find these conditions, we try to prove the theorem using variables X A, X B, Y B, X C, Y C, X A1, Y B1, Y C1 as parameters, i.e. allowing to multiply P with a polynomial in these variables. This succeeds and gives a coefficient c which has to be non zero. We take this c and factorise it (for example using MAPLE [4]). In this particular case, we get a product of 7 factors that we translate back as geometric conditions that are added to the goal as disjunctions. Finding these extra-conditions has to be done manually for the moment but with these extra conditions the theorem is proved automatically.

Trying to prove a general statement can give rise to extra conditions that have also to be proved. They correspond to denominators of all the fractions in the certificate being non-zero. When such a condition is not contradictory with the other hypotheses of the statement, it corresponds to an actual degenerated cases, so it is added to the original statement. Nevertheless, we have to be careful. Adding a condition that is contradictory with the other hypotheses would lead to a theorem that is trivially true but useless. For some examples, this detection of contradictions can be extremely costly.

Adding extra conditions has to be done for almost all theorems. This is not an easy task because selecting automatically which variables have to be put as parameters is not direct. As a matter of fact, there is a general method that we outline now. The set of coefficients c such that cP is in the ideal generated by $P_1, \ldots, P_s$ is itself an ideal. It defines an algebraic variety which represents the cases were the theorem is false. In order to completely describe this variety, one should compute its irreducible components. This can be done again with Gröbner bases computations. With the algebraic description of the variety of non-degenerate conditions, we will then be able to state the theorem correctly, even if some conditions are not easily expressible with the usual geometric predicates. For the theorems we have addressed so far, the heuristic method succeeds reasonably quickly, so we did not have to use the general machinery of irreducible decomposition of algebraic varieties yet.

We have compared our method with two systems: MACAULAY2 [10], a system dedicated to algebraic geometry which is very efficient in Gröbner bases computations and HOL LIGHT [14], a proof assistant that has a tactic for geometry theorem proving using Gröbner bases computation [15] [18] . For MACAULAY2, in several cases, e.g. Pascal's theorem for the circle, it was not able to check ideal membership (time over $1000s$) because it fails to compute the whole Gröbner basis while our method succeeds (in 397 seconds). For HOL LIGHT, Figure 4 gives a more extensive comparison. All theorems are general instances of the ones presented in Figure 2. Since our version of HOL LIGHT was running in interpreted mode, times have been divided by a factor of 4 to compensate the fact that COQ is using native code (4 is the average ratio between interpreted versus compiled code in OCAML). In HOL LIGHT, geometrical theorems are proven by refutation, i.e. the conclusion of the theorem is negated and the contradiction $1 = 0$ is proved by showing that 1 is in the Gröbner basis. Since this method differs from ours (the benefit of dividing the conclusion by the partial Gröbner basis during completion is lost), each theorem with hypothesis H, generic case C and particular cases $CP_1, \ldots, CP_n$ is proved using two equivalent formulations:

(1) the particular cases are negated in hypotheses: $H \wedge \neg CP_1 \wedge \ldots \wedge \neg CP_n \Rightarrow C$
(2) the conclusion is negated: $H \wedge \neg CP_1 \wedge \ldots \wedge \neg CP_n \wedge \neg C \Rightarrow 1 = 0$.

The idea is that the first version benefits from dividing the conclusion while the second one exactly mimics HOL LIGHT behaviour.

Let us now comment on these results:

- the first block of lines of the first table contains theorems for which refutation is slower than our method. For these theorems, our tactic is faster than HOL LIGHT.
- the second block contains theorems for which refutation is faster. For these theorems, our tactic and HOL LIGHT are similar.

	Times (seconds)			
	Coq (1)	HOL Light (1)	Coq (2)	HOL Light (2)
Feuerbach	94	>2000	>2000	>2000
Ptolemy	3.7	>800	20	>800
Ceva	4.8	28	5	28
Minh	0.8	1.9	27	1.9
Butterfly	24	20	25	21
Pappus	0.9	1	1	1
Euler circle	20	0.2	1	0.3
Pascal circle	50	6	11	6
Simson	89	118	7.3	121
Desargues	117	28	32	29
Threepoints	3	75	2.4	75

	Times (seconds)	
	Coq	HOL Light
Feuerbach or	1127	>2000
Ceva or	26	27
Pappus or	51	0.2
Desargues or	>500	11
Pascal circle or	44	8

Fig. 4. Comparison with HOL Light

- the second table contains theorems written as $H \Rightarrow C \vee CP_1 \vee \ldots \vee CP_n$. In this case, our method behaves very badly: the polynomial representing the conclusion is rather big, so dividing it to 0 takes a long time.

There is no clear winner between our method and the refutation one. However it seems that when the theorem deals with euclidean geometry and not only projective one, our method is better. In average, our tactic is faster than the one in HOL Light and it has the extra benefit of generating a certificate that can be easily verified.

6   Conclusion

This paper addresses an issue that is rarely taken into account by the auto-mated theorem proving community: can we really trust the tools we are using to prove theorems? For proof assistants, this question is central. Proof assistants are systems where mathematical knowledge is added progressively: new facts are derived from previously proved ones. Current systems usually come with li-braries that contain thousands of theorems. It is then crucial for these libraries to be built with the highest degree of confidence.

Nullstellensatz theorem and Gröbner bases algorithms are well-known ingre-dients to automatically prove theorems but how can they be put into action if, for example, one would like to build a library of geometrical facts that can be easily certified even by other proof assistants? The main contribution of this pa-per is to propose an effective way to do this. It is very easy to device a solution that works only on small examples. Before what is proposed in this paper, we had several non-conclusive attempts:

1. First, we have developed a certified implementation of Buchberger's algorithm that can be run within Coq [27]. This implementation was still an order of magnitude slower than usual implementations that can be found in computer algebra systems. Furthermore, any modification of this implementation usually requires a non-trivial proving effort to re-establish its correctness. Computations like the one needed for Ceva's theorem could not be performed inside Coq.

2. We have also tried to use efficient programs that computes Gröbner bases [8] using standard techniques of effective algebra [25]. These techniques make it possible to use the program as a black box but requires to add extra variables to the problem in order to compute the Q_i. The complexity of Gröbner bases being very sensitive to the number of variables, examples like Pascal were clearly out of reach.

3. We have also experimented with the simple form of certificates that only contains the coefficients of the linear combination as in the example (1) of page 47. Unfortunately, for some examples like Feuerbach, checking the certificates was taking much more time than generating them.

The notion of certificate as straight-line programs is a key aspect of this work. We agree that the only insight we could get is the fib example that shows that straightline program captures some cancellation. Still, we could not characterise when this actually happens. But it gives an explicit interface between the computation that is done externally of the proof assistant and the proof checking that is done inside the proof assistant. It also provides a very compact way of writing certificates. This means for example that transferring all the theorems of Table 2 into another proof assistant is straightforward: one just needs to develop his own trusted version of the checker. Note that in that case the time for generating the certificate becomes irrelevant: having Ceva's theorem would only require the time to check the certificate, i.e. a couple of seconds.

The system we have developed has made it possible to get very quickly a library of standard theorems of geometry within Coq. For building the library, we intensively use our secure decision procedure for ideal membership. Each theorem was first stated in its full generality. Then, our tactic that turns the problem in a membership problem and calls the decision procedure was tried. Most of the time, Coq failed to fully accept the certificate returned by the decision procedure. Some polynomials have to be proved non-zero. So we add them as extra conditions. As explained before, this was done manually. Once the extra conditions added, the tactic was tried again and this time the certificate was accepted. The theorem with its extra-conditions could then be stored in the Coq database. What was important in this experiment was to show that our decision procedure that is much slower than state-of-the-art Gröbner implementations could still be used for proving interesting theorems. We have encountered very few examples (Pascal's theorem for conics is one of them) where our decision procedure fails for a clear lack of computing power. This method of proving is of course very sensitive to the way theorems are stated. This is well known. Only one theorem has resisted our attempts to turn it into a COQ theorem, it is Morley's theorem (this theorem is in [5] but its statement involved extra points).

Looking at the problem of testing ideal membership from the perspective of generating small certificates is also very intriguing. We plan to further work on our generator. We believe it is possible to greatly improve both the efficiency and the compactness of the certificates that are generated. We also plan to apply similar ideas of compact certificates to other techniques that are used in geometry theorem proving. In that respect, Ritt-Wu's decomposition algorithm [6] seems a natural candidate.

Acknowledgements

This work was supported by the ANR Galapagos. Yves Bertot helped us with his expertise of the GEOGEBRA system [19]. We thanks the referees for their constructive comments on the first version of this paper.

References

 1. Allen, S.F., Constable, R.L., Howe, D.J., Aitken, W.E.: The Semantics of Reflected Proof. In: LICS, pp. 95–105. IEEE Computer Society, Los Alamitos (1990)
 2. Buchberger, B.: Bruno Buchberger's PhD thesis 1965: An algorithm for finding the basis elements of the residue class ring of a zero dimensional polynomial ideal. Journal of Symbolic Computation 41(3-4) (2006)
 3. Chaieb, A., Wenzel, M.: Context aware calculation and deduction. In: Kauers, M., Kerber, M., Miner, R., Windsteiger, W. (eds.) MKM/CALCULEMUS 2007. LNCS (LNAI), vol. 4573, pp. 27–39. Springer, Heidelberg (2007)
 4. Char, B.W., Fee, G.J., Geddes, K.O., Gonnet, G.H., Monagan, M.B.: A Tutorial Introduction to MAPLE. Journal of Symbolic Computation 2(2), 179–200 (1986)
 5. Chou, S.-C.: Mechanical geometry theorem proving. Kluwer Academic Publishers, Dordrecht (1987)
 6. Chou, S.-C., Gao, X.-S.: Ritt-Wu's Decomposition Algorithm and Geometry Theorem Proving. In: Stickel, M.E. (ed.) CADE 1990. LNCS, vol. 449, pp. 207–220. Springer, Heidelberg (1990)
 7. Eisenbud, D.: Commutative Algebra: with a View Toward Algebraic Geometry. Graduate Texts in Mathematics. Springer, Heidelberg (1999)
 8. Faugère, J.C.: A new efficient algorithm for computing Gröbner bases (f4). Journal of Pure and Applied Algebra 139(1/3), 61–88 (1999)
 9. Giusti, M., Heintz, J., Morais, J.E., Morgenstern, J., Pardo, L.M.: Straight-line programs in geometric elimination theory. Journal of Pure and Applied Algebra 124(1/3), 101–146 (1998)
10. Grayson, D.R., Stillman, M.E.: Macaulay2, http://www.math.uiuc.edu/Macaulay2/
11. Grégoire, B., Leroy, X.: A compiled implementation of strong reduction. In: International Conference on Functional Programming 2002, pp. 235–246. ACM Press, New York (2002)

12. Grégoire, B., Mahboubi, A.: Proving equalities in a commutative ring done right in coq. In: Hurd, J., Melham, T. (eds.) TPHOLs 2005. LNCS, vol. 3603, pp. 98–113. Springer, Heidelberg (2005)
13. Grégoire, B., Théry, L., Werner, B.: A computational approach to pocklington certificates in type theory. In: Hagiya, M. (ed.) FLOPS 2006. LNCS, vol. 3945, pp. 97–113. Springer, Heidelberg (2006)
14. Harrison, J.: HOL Light: A tutorial introduction. In: Srivas, M., Camilleri, A. (eds.) FMCAD 1996. LNCS, vol. 1166, pp. 265–269. Springer, Heidelberg (1996)
15. Harrison, J.: Complex quantifier elimination in HOL. In: TPHOLs 2001: Supplemental Proceedings. Division of Informatics, pp. 159–174. University of Edinburgh (2001), published as Informatics Report Series EDI-INF-RR-0046
16. Harrison, J.: Automating elementary number-theoretic proofs using gröbner bases. In: Pfenning, F. (ed.) CADE 2007. LNCS (LNAI), vol. 4603, pp. 51–66. Springer, Heidelberg (2007)
17. Harrison, J.: Verifying nonlinear real formulas via sums of squares. In: Schneider, K., Brandt, J. (eds.) TPHOLs 2007. LNCS, vol. 4732, pp. 102–118. Springer, Heidelberg (2007)
18. Harrison, J.: Handbook of Practical Logic and Automated Reasoning. Cambridge University Press, Cambridge (2009)
19. Hohenwarter, M., Preiner, J.: Dynamic Mathematics with GeoGebra. Journal of Online Mathematics 7, ID 1448 (March 2007)
20. Kapur, D.: Geometry theorem proving using Hilbert's Nullstellensatz. In: SYMSAC 1986: Proceedings of the Fifth ACM Symposium on Symbolic and Algebraic Computation, pp. 202–208. ACM, New York (1986)
21. Kapur, D.: A refutational approach to geometry theorem proving. Artificial Intelligence 37(1-3), 61–93 (1988)
22. Kapur, D.: Automated Geometric Reasoning: Dixon Resultants, Gröbner Bases, and Characteristic Sets. In: Wang, D. (ed.) ADG 1996. LNCS, vol. 1360, pp. 1–36. Springer, Heidelberg (1998)
23. Kreisel, G., Krivine, J.L.: Elements of Mathematical Logic (Model Theory). Studies in Logic and the Foundations of Mathematics. North-Holland, Amsterdam (1967)
24. Paulson, L.C.: Isabelle: A generic theorem prover. Journal of Automated Reasoning 828 (1994)
25. Pottier, L.: Connecting Gröbner Bases Programs with Coq to do Proofs in Algebra, Geometry and Arithmetics. In: Proceedings of the LPAR Workshops: Knowledge Exchange: Automated Provers and Proof Assistants, and The 7th International Workshop on the Implementation of Logics. CEUR Workshop Proceedings, vol. (418) (2008)
26. Robu, J.: Geometry Theorem Proving in the Frame of the Theorema Project. Tech. Rep. 02-23, RISC Report Series, University of Linz, Austria, phD Thesis (September 2002)
27. Théry, L.: A Machine-Checked Implementation of Buchberger's Algorithm. Journal of Automated Reasoning 26(2) (2001)
28. Wiedijk, F.: Formalizing 100 Theorems, http://www.cs.ru.nl/~freek/100
29. Wu, W.-T.: On the Decision Problem and the Mechanization of Theorem-Proving in Elementary Geometry. In: Automated Theorem Proving - After 25 Years, pp. 213–234. American Mathematical Society, Providence (1984)

Multivariate Resultants in Bernstein Basis

Deepak Kapur* and Manfred Minimair

[1] University of New Mexico, Department of Computer Science, Albuquerque,
New Mexico 87131, USA
kapur@cs.unm.edu
[2] Seton Hall University, Department of Mathematics and Computer Science,
400 South Orange Avenue, South Orange, New Jersey 07079, USA
manfred@minimair.org

Abstract. Macaulay and Dixon resultant formulations are proposed for parametrized multivariate polynomial systems represented in Bernstein basis. It is proved that the Macaulay resultant for a polynomial system in Bernstein basis vanishes for the total degree case if and only if the either the polynomial system has a common Bernstein-toric root, a common infinite root, or the leading forms of the polynomial system obtained by replacing every variable x_i in the original polynomial system by $\frac{y_i}{1+y_i}$ have a non-trivial common root. For the Dixon resultant formulation, the rank sub-matrix constructions for the original system and the transformed system are shown to be essentially equivalent. Known results about exactness of Dixon resultants of a sub-class of polynomial systems as discussed in Chtcherba and Kapur in Journal of Symbolic Computation (August, 2003) carry over to polynomial systems represented in the Bernstein basis. Furthermore, in certain cases, when the extraneous factor in a projection operator constructed from the Dixon resultant formulation is precisely known, such results also carry over to projection operators of polynomial systems in the Bernstein basis where extraneous factors are precisely known. Applications of these results in the context of geometry theorem proving, implicitization and intersection of surfaces with curves are discussed. While Macaulay matrices become large when polynomials in Bernstein bases are used for problems in these applications, Dixon matrices are roughly of the same size.

Keywords: resultants, Macaulay matrix, Dixon matrix, Bernstein basis.

1 Introduction

Resultants provide a necessary and sufficient condition on a polynomial system to have a common root. If a polynomial system is parametrized, then the resultant is a polynomial in the parameters such that the resultant polynomial vanishes for parameter values if and only if the instantiated polynomial system has a common solution (in a suitable space). Typically resultant computations are defined for polynomials represented in the standard power basis. However,

---

* Partially supported by NSF Award CCF-0729097.

T. Sturm and C. Zengler (Eds.): ADG 2008, LNAI 6301, pp. 60–85, 2011.

there are applications such as computer-aided design, graphics, robotics, chemical kinematics, for which bases other than the standard power basis are more suitable. A case in point is the Bernstein basis representation which has been used extensively in computer-aided design [22] and also in constructive approximation theory [19].

Fundamental operations for polynomials represented with respect to bases other than the usual power basis are being intensely studied. Examples include computation of resultants and resultant matrices for univariate polynomials [29,36,37,41,7,40], gcds [31,20,11], generalized companion matrices [2,38,39,16], polynomial remainder sequences [4,23], and polynomial division [3,35,1,33]. Carrying out fundamental operations over alternative bases is motivated by the desire to avoid computational cost and numeric errors incurred by converting between different polynomial bases.

We show in this paper that multivariate resultant formulations, including Macaulay resultants [17] and Dixon resultants [26], can be adapted in a natural way to be applicable to polynomials represented in the Bernstein basis. This work is in part motivated by [29,41,7,40] that consider univariate cases. It is also motivated by [30] that studies polynomial systems arising from surface-curve intersections in Bernstein basis and constructs resultant matrices for these special systems. However, [30] neither studies any properties of these resultant matrices, such as Theorems 16, 21, and 22 of the current paper, nor considers the general case of arbitrary polynomial systems as it is done in the current paper.

The Macaulay resultant ("total-degree resultant") for a polynomial system with x_i's as its variables and represented in the Bernstein basis with total degree support can be defined by doing a transformation of the polynomials in the Bernstein basis into a related polynomial system in the standard power basis using a different set of variables; every variable x_i in the original polynomial system is replaced by $\frac{y_i}{1+y_i}$. While preserving the original coefficients, the transformed polynomial system is shown to be in the standard power basis. So the classical Macaulay construction applies, however, leading to a Macaulay matrix with respect to the Bernstein basis. The main result about the Macaulay resultant formulation for a polynomial system in Bernstein basis is that the total degree resultant of the polynomial system vanishes if and only if the polynomial system has a common Bernstein-toric root (defined precisely later), a common infinite root, or the leading forms of the transformed polynomial system have a common (nontrivial) root.

In the case of the Dixon resultant formulation, the results are even more direct. For polynomials represented in variables x_i's with unmixed Bernstein basis degree, replacement of each x_i by $\frac{y_i}{1-y_i}$, where the y_i's are new variables, leads to a relationship between the Dixon matrix of the original system and the Dixon matrix of the transformed system. Because of this, known results about exactness of Dixon resultants of a sub-class of polynomial systems as discussed in [12] carry over to polynomial systems represented in the Bernstein basis. Furthermore, in certain cases, when the extraneous factor in a projection operator constructed from the Dixon resultant formulation is precisely known, the extraneous factor in the projection operator of a polynomial system in the Bernstein basis can be predicted a priori.

For polynomial systems represented using the Bernstein basis with mixed basis degrees, we observe that such bases can be lifted to unmixed basis degree, so that the results of the previous section apply. Lifting results in larger matrices because of which additional extraneous factors are introduced into the projection operators computed for the lifted unmixed polynomial systems represented in the Bernstein basis. Thus we present an algorithm to reduce the Dixon matrix to minimal size, eliminating the additional factors.

Furthermore, we show that the rank sub-matrix construction of Kapur, Saxena and Yang [26], can be used to extract the resultant from the projection operator generated by the Dixon matrix in Bernstein basis representation in the same way as for power basis representation. This construction is especially useful when polynomial systems are not generic and/or parameters are specialized. Results from [25] where it is shown that Dixon resultant formulation exploits the sparse structure and specialization of coefficients of terms in polynomial systems, should extend as well, we believe, to specialized polynomial systems in the Bernstein basis.

As already mentioned, polynomials represented in Bernstein basis have important applications in computer graphics and in constructive approximation theory. Let us elaborate and discuss the need for resultants for Bernstein polynomials. In computer graphics Bernstein polynomials are used to represent Bézier curves and surfaces. For example a surface is given by three bivariate polynomials of the form

$$\sum_{i_1 i_2} a_{k\, i_1 i_2}\, \beta_{(n_{k1}, n_{k2}), (i_1, i_2)}(x_1, x_2),$$

for $k = 0, 1, 2$, where the tuples $(a_{0\, i_1 i_2}, a_{1\, i_1 i_2}, a_{2\, i_1 i_2})$ represent the so-called control points for the surface. (The β's represent Bernstein basis elements. See Sect. 2.4.) In approximation theory Bernstein polynomials are used to approximate continuous functions. In both areas polynomial equations in Bernstein basis naturally arise, for example when intersecting curves and surfaces or when computing roots. Obviously resultants can be used to study such systems of equations. The techniques presented in the current paper can be used for such studies *directly* without converting the polynomials into power basis representation. This provides practical benefits by avoiding the computational cost of such a conversion. Moreover, for the case of numeric coefficients such a conversion is normally not recommended because of numerical instability and possible severe loss of accuracy incurred by the conversion. Applications of these results in the context of geometry theorem proving, implicitization and intersection of surfaces with curves are discussed. While Macaulay matrices become large when polynomials in Bernstein bases are used for problems in these applications, Dixon matrices are roughly of the same size.

The paper is organized as follows. Section 2 reviews background material on multivariate resultants and projection operators, particularly Macaulay and Cayley-Dixon resultant formulations. Section 2.4 is on multivariate Bernstein bases for polynomials. Section 3 discusses how the Macaulay resultant formulation can be adapted to work on polynomial systems represented in Bernstein bases. This is followed by the Cayley-Dixon resultant formulation for polynomial systems using Bernstein bases (Sect. 4). The next Sect. 5 discusses some applications. The paper ends with a section on conclusion and future work. The appendix includes the proofs of the main results in the paper.

2   Preliminaries

2.1   Multivariate Resultants and Projection Operators

We summarize various notions of resultants and projection operators considered in this paper and elaborate on their constructions in the corresponding sections below.

By the resultant of a multi-variate polynomial system $F = (f_0, \ldots, f_l)$ one usually means a polynomial in the coefficients of the f_i's that vanishes if and only if the f_i's have a common root in a given variety. For the cases of projective and general toric varieties one can use a Macaulay-style approach [17,18] for constructing the resultant.

By the projection operator of a multi-variate polynomial system F one usually means a *necessary* condition for a common root of F, that is, a polynomial in the coefficients of the f_i's that vanishes if the f_i's have a common root in a given variety. Such projection operators can be constructed following a Bézout/Cayley/Dixon-style approach [21,26]. It has been shown that under some natural assumptions [9,15] these projection operators are multiples of resultants defined over suitable varieties.

2.2   Macaulay-Style Approach

The Macaulay-style approach constructs resultants for a multi-variate polynomial system $F = (f_0, \ldots, f_l)$ in the variables $X = (x_1, \ldots, x_l)$ as the quotient of two determinants [28,10,18]. The matrix in the numerator consists of the coefficients of suitable multiples of the f_j's. The multipliers for the f_j's are power products of the variables in X. The matrix in the denominator is a suitably chosen sub-matrix of the matrix in the numerator. This construction leads to resultants defined over the projective space as well as over toric varieties in general. Section 3 will study this construction for polynomials presented in Bernstein basis.

2.3   Bézout/Cayley/Dixon-Style Approach

The Bézout/Cayley/Dixon-style approach also constructs a matrix from multiples of the polynomials in the system F as above [21]. However, unlike the Macaulay-style approach, the multipliers usually are polynomials instead of monomials. This approach usually leads to a smaller matrix and thus is faster in computations. The resultants extracted from this matrix are defined over suitably parametrized varieties (including toric varieties) [9,26] (or affine in certain cases). Section 4 will study this construction for polynomials presented in Bernstein basis.

2.4 Multivariate (Tensor-Product) Bernstein Bases

We review the standard notation for Bernstein bases (see e.g. [5]). Note that in the literature Bernstein bases for multivariate polynomials are also called tensor-product Bernstein bases because they are obtained as products of univariate Bernstein bases, similarly to multivariate power bases.

Let $N = (n_1, \ldots, n_l)$ and $I = (i_1, \ldots, i_l)$ be multi-indices such that $I \leq N$, that is, $i_j \leq n_j$, for all j. Furthermore, let $X = (x_1, \ldots, x_l)$ be a list of variables x_j. Analogous to a power product $x_1^{i_1} \cdots x_l^{i_l}$, we define the I-th Bernstein basis element (of degree N) as

$$\beta_{N,I}(X) := \binom{n_1}{i_1} \cdots \binom{n_l}{i_l} x_1^{i_1} (1 - x_1)^{n_1 - i_1} \cdots x_l^{i_l} (1 - x_l)^{n_l - i_l},$$

where $\binom{a}{b}$ is the usual binomial coefficient. Furthermore note that often we will abbreviate the product $\binom{n_1}{i_1} \cdots \binom{n_l}{i_l}$ in the definition of $\beta_{N,I}(X)$ with $\binom{N}{I}$. Furthermore, we call the products $x_1^{i_1} (1 - x_1)^{n_1 - i_1} \cdots x_l^{i_l} (1 - x_l)^{n_l - i_l}$ without binomial coefficients the I-th *scaled* Bernstein basis element.

As is well-known [5], all polynomials (over some field k) in variables X, whose degree in each x_j is less than or equal to n_j, can be written as linear combinations of Bernstein basis elements $\beta_{N,I}(X)$ with $I \leq N$.

Example 1. Consider the polynomial

$$f = a_{21} x_1^2 x_2 + a_{20} x_1^2 (1 - x_2) + 2 a_{11} x_1 (1 - x_1) x_2$$
$$+ 2 a_{10} x_1 (1 - x_1) (1 - x_2) + a_{01} (1 - x_1)^2 x_2$$
$$+ a_{00} (1 - x_1)^2 (1 - x_2)$$

in the variables x_1 and x_2. Using the standard notation, we have

$$f = a_{21} \beta_{N,(2,1)}(X) + a_{20} \beta_{N,(2,0)}(X) + a_{11} \beta_{N,(1,1)}(X)$$
$$+ a_{10} \beta_{N,(1,0)}(X) + a_{01} \beta_{N,(0,1)}(X)$$
$$+ a_{00} \beta_{N,(0,0)}(X),$$

where $N = (2, 1)$ is the basis degree and $X = (x_1, x_2)$ are the variables of f.

3 An Analogue of Macaulay-Style Resultant Construction

This section constructs analogues of Macaulay-style resultants for polynomials with total-degree support with respect to the Bernstein basis. The notion of support with respect to the Bernstein basis is defined as for power basis.

Definition 2 (Bernstein basis support). *The support of the polynomial*

$$\sum_{I \in S_N} a_I \, \beta_{N,I}(X)$$

(with respect to the Bernstein basis of degree N) is the set S_N, a subset of all multi-indices $I \leq N$. (We use the subscript N for the support set S in order to emphasize that the support depends on the basis-degree N.)

The construction proposed in this section is a multi-variate generalization of [41] for univariate polynomials. In [41], the authors define the univariate resultant as the determinant of a modified Sylvester matrix where the matrix entries are the coefficients of the Bernstein polynomials multiplied by binomial coefficients corresponding to their Bernstein basis elements. The major difficulty in generalizing this approach comes from the observation that multi-variate Bernstein polynomials with total-degree supports have the (generic) root $(1, \ldots, 1)$ as we will see below in Example 7.

The following definition transfers the notion of total-degree support in terms of the power basis to the Bernstein basis. The essential property of total-degree support is a geometric one. That is, the support set contains all multi-indices whose absolute value is less than or equal to some given bound. The convex hull of such a set is called a simplex. Therefore it would be quite natural to use the term "simplex support" instead of total-degree support. But, since we want to follow the standard notation in power basis as much as possible we keep the usual term of total-degree support.

Definition 3. *Let S_N be the Bernstein basis support of the polynomial f as in Definition 2. Then S_N is called the total-degree m support (with respect to the Bernstein basis of degree N) iff S_N consists of all multi-indices I with $|I| := i_1 + \cdots + i_l \leq m$. Furthermore, we say f has total-degree m support (with respect to the Bernstein basis of degree N).*

Remark 4. Even though the degree N of the Bernstein basis is not directly related to the bound m, the definition of total-degree Bernstein basis support implies certain inequalities. One important inequality is $m \leq n_j$. Otherwise the support S_N could not contain *all* multi-indices I with $|I| \leq m$ such as $(m, 0, \ldots, 0)$ which corresponds to the basis element $\beta_{N,(m,0,\ldots,0)}(X)$. Another inequality follows from this, namely, $m < |N| = n_1 + n_2 + \cdots + n_l$ if $1 < l$ and $1 \leq m$.

See the Example 5 below where the polynomial f has Bernstein basis degree $N = (3, 2)$ but "total-degree 2 support".

Example 5. Consider the polynomial

$$f = a_{20} \, \beta_{N,(2,0)}(X) + a_{11} \, \beta_{N,(1,1)}(X) + a_{02} \, \beta_{N,(0,2)}(X)$$
$$+ a_{10} \, \beta_{N,(1,0)}(X) + a_{01} \, \beta_{N,(0,1)}(X) + a_{00} \, \beta_{N,(0,0)}(X),$$

where $N = (n_1, n_2) = (3, 2)$ and $X = (x_1, x_2)$. Then f has total-degree 2 support with respect to the Bernstein basis.

As already noted, this definition is analogous to the definition of total-degree support with respect to the usual power basis. But there are significant differences of which we illustrate two.

First, Bernstein polynomials with total-degree support do not have total-degree support with respect to the power basis.

Example 6 (Example 5 continued).
After converting into power-basis representation the polynomial f has support

$$\{(3,2),(3,1),(3,0),(2,2),(2,1),(2,0),(1,2),(1,1),(1,0),(0,2),(0,1),(0,0)\}$$

with respect to the power basis which is not a total-degree support. The total degree of f with respect to power basis is even higher than 2. It is 5 which would lead to a much larger resultant matrix than for degree 2.

Second, polynomials with total-degree Bernstein basis support vanish for the tuple $X = (1, \ldots, 1)$.

Example 7 (Example 6 continued). Note that f does not contain the Bernstein basis element $\beta_{N,(3,2)}(X) = x_1^3 x_2^2$ and all the other basis elements either contain the factor $(1 - x_1)$ or the factor $(1 - x_2)$. Thus f vanishes for $X = (1, 1)$.

The property illustrated in the previous example can be shown for general total-degree Bernstein basis support as stated in the following theorem.

Theorem 8. *Let f be a polynomial in the variables $X = (x_1, \ldots, x_l)$, where $l \geq 2$, with total-degree m support with respect to the Bernstein basis and with $m \geq 1$. Then f has the root $X = (1, \ldots, 1)$.*

Therefore we have to define the total-degree resultant such that the root $(1, \ldots, 1)$ is ignored. Towards this goal, we will show how to construct another system of polynomials from a given one. This constructed system preserves the essential common roots of the original one. Thus the resultant of the original system can be defined as the resultant of the newly constructed one.

Before stating the definition formally we give a motivating example.

Example 9. Consider the polynomials

$$f_j = a_{j20}\, x_1^2\, (1 - x_2)^2 + 4\, a_{j11}\, x_1\, (1 - x_1)\, x_2\, (1 - x_2) + a_{j02}\, (1 - x_1)^2\, x_2^2$$
$$+ 2\, a_{j10}\, x_1\, (1 - x_1)\, (1 - x_2)^2 + 2\, a_{j01}\, (1 - x_1)^2\, x_2\, (1 - x_2)$$
$$+ a_{j00}\, (1 - x_1)^2\, (1 - x_2)^2$$

with total-degree 2 support with respect to the Bernstein basis for $j = 0, 1, 2$. (Note that the total degree of f with respect to the *power basis* is indeed 4.) From the f_j's we construct polynomials h_j by substituting $x_i = 1 + z_i$, that is,

$$h_j = f_j(1 + z_1, 1 + z_2)$$
$$= a_{j20}\, (1 + z_1)^2\, z_2^2 + 4\, a_{j11}\, (1 + z_1)\, z_1\, (1 + z_2)\, z_2 + a_{j02}\, z_1^2\, (1 + z_2)^2$$
$$- 2\, a_{j10}\, (1 + z_1)\, z_1\, z_2^2 - 2\, a_{j01}\, z_1^2\, (1 + z_2)\, z_2$$
$$+ a_{j00}\, z_1^2\, z_2^2$$

Thus $h_j(0,0) = f_j(1,1) = 0$. Obviously, the substitution $x_i = z_i + 1$ bijectively transforms the roots of f_j into the roots of h_j. Therefore one can define the resultant of the f_j's as the *toric* resultant [17] of the h_j's. Since the toric resultant ignores roots with $z_1 = 0$ or $z_2 = 0$ it does not trivially vanish for the h_j's and therefore it is well-defined. Still, it is possible to simplify this definition of the resultant of the f_j's. That is, we will illustrate a transformation which maps h_j into certain g_j's (from Definition 13) under which the toric resultant can be shown to be invariant. Thus we define the resultant of the f_j's to be the toric resultant of the g_j's, which by definition is the Macaulay (projective) resultant of the g_j's.

In order to construct the g_j's we substitute $z_i = (-y_i - 1)^{-1}$ into h_j and multiply the result by $(-y_1 - 1)^2(-y_2 - 1)^2$, that is,

$$
\begin{aligned}
g_j &= (-y_1 - 1)^2 (-y_2 - 1)^2 \, h_j((-y_1 - 1)^{-1}, (-y_2 - 1)^{-1}) \\
&= (-y_1 - 1)^2 (-y_2 - 1)^2 \left(a_{j20} \left(\frac{-y_1}{-y_1 - 1} \right)^2 (-y_2 - 1)^{-2} \right. \\
&\quad + 4\, a_{j11} \frac{-y_1}{-y_1 - 1} (-y_1 - 1)^{-1} \frac{-y_2}{-y_2 - 1} (-y_2 - 1)^{-1} \\
&\quad + a_{j02} (-y_1 - 1)^{-2} \left(\frac{-y_2}{-y_2 - 1} \right)^2 \\
&\quad - 2\, a_{j10} \frac{-y_1}{-y_1 - 1} (-y_1 - 1)^{-1} (-y_2 - 1)^{-2} \\
&\quad - 2\, a_{j01} (-y_1 - 1)^{-2} \frac{-y_2}{-y_2 - 1} (-y_2 - 1)^{-1} \\
&\quad \left. + a_{j00} (-y_1 - 1)^{-2} (-y_2 - 1)^{-2} \right) \\
&= a_{j20} \, y_1^2 + 4\, a_{j11} \, y_1 y_2 + a_{j02} \, y_2^2 + 2\, a_{j10} \, y_1 + 2\, a_{j01} \, y_2 + a_{j00}.
\end{aligned}
$$

Thus the vanishing of the resultant of the f_j's is equivalent to either the f_j's having a common root with $x_i \neq 1$ (or possibly x_i at infinity), for all i, or the leading forms $a_{j20} \, y_1^2 + 4\, a_{j11} \, y_1 y_2 + a_{j02} \, y_2^2$ having a common root $\neq 0$ (compare Theorem 16).

Remark 10. As illustrated in the previous example, total-degree resultants for Bernstein polynomials will capture roots X where $x_i \neq 1$ for all i. Roots with some x_i's being 1 are not captured. If one wants to deal with such roots one can substitute 1 for the corresponding x_i's and thus obtain a strongly overdetermined system of equations with fewer variables. This situation is analogous to toric resultants which ignore roots where some x_i vanish.

Next we define the notions of Bernstein-toric, non-trivial and infinite root which will be used to describe the vanishing of the total-degree Bernstein resultant with respect to Bernstein basis. The name Bernstein-toric is motivated by the standard notion of toric. A toric root is a root X where no x_i vanishes, whereas a Bernstein-toric root is one where no x_i equals 1. Non-trivial root is a well-known notion.

Furthermore, infinite roots are defined particularly for total-degree Bernstein polynomials and are illustrated in the example following the formal definition.

Definition 11. *The root $\mathcal{X} = (\xi_1, \ldots, \xi_l)$ of the polynomial f is called*

1. *Bernstein-toric iff $\xi_i \neq 1$ for all i,*
2. *non-trivial iff $\xi_i \neq 0$ for one i, that is $\mathcal{X} \neq 0$.*

Furthermore the tuple $\mathcal{X}$ is called infinite root of f iff there is a strict subset of variables $X_k = (x_{i_1}, \ldots, x_{i_k})$ such that $\overline{\mathcal{X}} = (\ldots, \xi_i, \ldots)_{i \notin \{i_1, \ldots, i_k\}}$ is a root of $f^{[X_k]}$, where the polynomial $f^{[X_k]}$ is recursively defined as the leading coefficient, with respect to x_{i_k}, of $f^{[X_{k-1}]}$, and where $f^{[X_0]} := f$.

Example 12 (Infinite root for total-degree Bernstein basis support). Let

$$f = a_{20}\, x_1^2 \,(1 - x_2)^2 \,+\, 4\, a_{11}\, x_1 \,(1 - x_1)\, x_2 \,(1 - x_2) \,+\, a_{02} \,(1 - x_1)^2 \,x_2^2$$
$$+\, 2\, a_{10}\, x_1 \,(1 - x_1)\,(1 - x_2)^2 \,+\, 2\, a_{01} \,(1 - x_1)^2 \,x_2 \,(1 - x_2)$$
$$+\, a_{00} \,(1 - x_1)^2 \,(1 - x_2)^2.$$

f homogenized with respect to x_1 (with homogenizing variable z_1) is

$$f^{h_1} = a_{20}\, x_1^2 \,(1 - x_2)^2 \,+\, 4\, a_{11}\, x_1 \,(z_1 - x_1)\, x_2 \,(1 - x_2) \,+\, a_{02} \,(z_1 - x_1)^2 \,x_2^2$$
$$+\, 2\, a_{10}\, x_1 \,(z_1 - x_1)\,(1 - x_2)^2 \,+\, 2\, a_{01} \,(z_1 - x_1)^2 \,x_2 \,(1 - x_2)$$
$$+\, a_{00} \,(z_1 - x_1)^2 \,(1 - x_2)^2.$$

Thus the leading coefficient of x_1 is $f^{h_1}(1, 0)$, that is,

$$f^{[X_1]} = a_{20} \,(1 - x_2)^2 \,+\, 4\, a_{11} \,(-1)\, x_2 \,(1 - x_2) \,+\, a_{02} \,(-1)^2 \,x_2^2$$
$$+\, 2\, a_{10} \,(-1)\,(1 - x_2)^2 \,+\, 2\, a_{01} \,(-1)^2 \,x_2 \,(1 - x_2)$$
$$+\, a_{00} \,(-1)^2 \,(1 - x_2)^2,$$

where $X_1 = (x_1)$. Similarly, the leading coefficient of x_2 in $f^{[X_1]}$ is

$$f^{[X_2]} := a_{20} \,(-1)^2 \,+\, 4\, a_{11} \,(-1)\,(-1) \,+\, a_{02} \,(-1)^2$$
$$+\, 2\, a_{10} \,(-1)\,(-1)^2 \,+\, 2\, a_{01} \,(-1)^2 \,(-1)$$
$$+\, a_{00} \,(-1)^2 \,(-1)^2,$$

where $X_2 = (x_1, x_2)$. Then any root of $f^{[X_1]}$ or $f^{[X_2]}$ is called an infinite root of f.

Definition 13 (Total-degree Bernstein resultant). *Let*

$$f_j = \sum_{\substack{I \leq N_j, \\ |I| \leq m_j}} a_I \, \beta_{N_j, I}(X)$$

be polynomials with total-degree m_j support with respect to their Bernstein bases N_j, for $j = 0, \ldots, l$. Furthermore, let

$$g_j = \sum_{\substack{I \leq N_j, \\ |I| \leq m_j}} a_I \binom{N_j}{I} Y^I,$$

where $Y^I = y_1^{i_1} \cdots y_l^{i_l}$, for $j = 0, \ldots, l$. Then the total-degree Bernstein resultant of $f_0, \ldots, f_l$ is defined as the usual Macaulay (projective) resultant of $g_0, \ldots, g_l$ (with respect to the power basis in variables Y).

Remark 14. We found that the Macaulay matrices generated when computing total-degree Bernstein resultants can also be applied to polynomials whose supports are strict subsets of total-degree Bernstein supports. For details, see Sect. 5.3 which covers applications where such supports arise.

Remark 15. Using the terminology of [41] the Macaulay matrix of the resultant of the g_j's is called the Macaulay matrix of the f_j's with respect to the "scaled Bernstein basis". Analogous to [41] one can show that this matrix can be converted into a matrix with respect to the standard Bernstein basis by multiplying its columns by suitable constants. It is important to note that this conversion only changes the value of the determinant of the Macaulay matrix by a constant factor. Also note that in the univariate case with $m_0 = N_0$ and $m_1 = N_1$, the Macaulay matrix with respect to the scaled Bernstein basis agrees with the Sylvester matrix of [41] (up to permutations of rows and columns) because its sizes are $(m_0 + m_1) \times (m_0 + m_1)$.

The following theorem shows when the total-degree resultant vanishes.

Theorem 16. *The total-degree Bernstein resultant of $f_0, \ldots, f_l$ vanishes iff the polynomials $f_0, \ldots, f_l$ have a common*

1. *Bernstein-toric root, or*
2. *infinite root*

or the forms

$$\sum_{|I|=m_j} a_I \binom{N_j}{I} Y^I, \tag{1}$$

for $j = 0, \ldots, l$, have a common non-trivial root.

Obviously, one could also state the condition (1) in terms of the Bernstein basis rather than the power basis Y^I. However, the presentation in power basis is more compact and therefore it is given in the theorem.

4 Bézout/Cayley/Dixon-Style Resultant Construction

This section provides an approach for constructing the Dixon matrix [26] for multi-variate Bernstein polynomials. This construction will be carried out without converting the Bernstein polynomials into power basis representation. Furthermore, the constructed Dixon matrix will be given with respect to a suitable Bernstein basis rather than the usual power basis. The approach of this section is different from [6] which considers the case of two univariate polynomials. The approach parallels [41] in the sense that it constructs a matrix (Dixon matrix) with respect to the scaled Bernstein basis which can be converted into a matrix with respect to the Bernstein basis by adjusting matrix entries by constant factors.

This section is divided into three parts. The first part considers the construction for polynomials with unmixed Bernstein basis degrees (Definition 17). The second part extends the first part to polynomials with arbitrary (mixed) Bernstein basis degrees. The third part studies the rank sub-matrix construction for Dixon matrices with respect to Bernstein bases.

4.1 Unmixed Bernstein Basis Degrees

Definition 17. *Let* $F = (f_0, \ldots, f_l)$ *be a list of multivariate polynomials in variables* $X = (x_1, \ldots, x_l)$ *where* f_j *is represented in Bernstein bases of degrees* N_j. *Then*

1. *F has unmixed Bernstein basis degree iff* $N_1 = \cdots = N_l$, *and*
2. *F has mixed Bernstein basis degree iff the N_j's are not-necessarily equal.*

Example 18. Consider

$$f_0 = a_{22}\, \beta_{(2,3),(2,2)}(x_1, x_2) + a_{11}\, \beta_{(2,3),(1,1)}(x_1, x_2),$$
$$f_1 = b_{21}\, \beta_{(2,3),(2,1)}(x_1, x_2) + b_{02}\, \beta_{(2,3),(0,2)}(x_1, x_2).$$

Then f_0 and f_1 have unmixed Bernstein basis degree $(2, 3)$.

Remark 19. It is important not to confuse the notions of (un)mixed Bernstein basis degrees and (un)mixed Bernstein basis supports. Notice that f_0 and f_1 from Example 18 have unmixed Bernstein basis degree but they have mixed Bernstein basis support. Furthermore, notice that polynomials whose Bernstein basis degrees are mixed, necessarily have mixed Bernstein basis support.

Before giving the construction for the general case, we consider a simple univariate example.

Example 20. Let $F = (f_0, f_1)$ with

$$f_j = a_{j3}\, \beta_{3,3}(x) + a_{j2}\, \beta_{3,2}(x) + a_{j1}\, \beta_{3,1}(x) + a_{j0}\, \beta_{3,0}(x).$$

Observe that f_0 and f_1 have the same basis degree 3. By using the well-known substitution $x = \frac{y}{1+y}$ we can write $f_j = (1-x)^3 g_j(\frac{x}{1-x})$ where

$$g_j = a_{j3}\, y^3 + 3\, a_{j2}\, y^2 + 3\, a_{j1}\, y + a_{j0}.$$

Next let us construct the Dixon (Bézout) polynomial θ_F of f_0 and f_1. We get

$$\theta_F = \begin{vmatrix} f_0(x) & f_1(x) \\ f_0(\overline{x}) & f_1(\overline{x}) \end{vmatrix} \Big/ (x - \overline{x}),$$

where the notation $f_j(x)$ means that f_j is considered as a polynomial in x, as it is given, and $f_j(\overline{x})$ means the polynomial obtained from $f_j(x)$ by replacing x with $\overline{x}$. By substituting the g_j's we obtain

$$\theta_F = \begin{vmatrix} (1-x)^3\, g_0(y) & (1-x)^3\, g_1(y) \\ (1-\overline{x})^3\, g_0(\overline{y}) & (1-\overline{x})^3\, g_1(\overline{y}) \end{vmatrix} \Big/ (x-\overline{x})$$

$$= \begin{vmatrix} g_0(y) & g_1(y) \\ g_0(\overline{y}) & g_1(\overline{y}) \end{vmatrix} \frac{(1-\overline{x})^3 (1-x)^3}{x - \overline{x}},$$

where $\overline{y} = \frac{\overline{x}}{1-\overline{x}}$. Notice that $\frac{(1-\overline{x})^3(1-x)^3}{x-\overline{x}} = \frac{(1-\overline{x})^2(1-x)^2}{y-\overline{y}}$. Thus

$$\theta_F = (1-\overline{x})^2\, (1-x)^2\, \theta_G(\frac{\overline{x}}{1-\overline{x}}, \frac{x}{1-x}), \qquad (2)$$

where θ_G is viewed as a polynomial in $y, \overline{y}$.

Observe that the right-hand side of (2) represents the Dixon polynomial of F as a linear combination of suitable power products $\overline{x}^{\overline{i}}\, (1-\overline{x})^{2-\overline{i}}\, x^i\, (1-x)^{2-i}$. Thus the matrix obtained by viewing the right-hand side as a bilinear form in the $\overline{x}^{\overline{i}}\, (1-\overline{x})^{2-\overline{i}}$'s and the $x^i\, (1-x)^{2-i}$'s is the Dixon matrix of F with respect to the scaled Bernstein basis. By adjusting rows and columns by corresponding factors $\left(\frac{2}{\overline{i}}\right)^{-1}$ and $\left(\frac{2}{i}\right)^{-1}$ one obtains the Dixon matrix of F with respect to the Bernstein basis.

The following theorems describe, in general, the construction of the Dixon polynomial and matrix for Bernstein polynomials.

Theorem 21. *Let $F = (f_0, \ldots, f_l)$ be a list of polynomials in variables $X = (x_1, \ldots, x_l)$ such that all f_j's have the same (unmixed) Bernstein basis degree $N = (n_0, \ldots, n_l)$. Furthermore, let g_j be a polynomial in the variables $Y = (y_1, \ldots, y_l)$ such that*

$$f_j = (1-x_1)^{n_1} \ldots (1-x_l)^{n_l}\, g_j(\frac{x_1}{1-x_1}, \ldots, \frac{x_l}{1-x_l})$$

and let $G = (g_0, \ldots, g_l)$. Then the coefficient of $\beta_{N', \overline{I}}(\overline{X}) \cdot \beta_{N', I}(X)$ in the Dixon polynomial of F is $\left(\frac{\overline{N}'}{\overline{I}}\right)^{-1} \cdot \left(\frac{N'}{I}\right)^{-1}$ times the coefficient of $\overline{Y}^{\overline{I}}\, Y^I$ of the Dixon polynomial of G, where $\overline{N}' = (l\, n_1 - 1, (l-1)\, n_2 - 1, \ldots, n_l - 1)$, $N' = (n_1 - 1, 2\, n_2 - 1, \ldots, l\, n_l - 1)$, $\overline{X} = (\overline{x}_1, \ldots, \overline{x}_l)$ and $\overline{Y} = (\overline{y}_1, \ldots, \overline{y}_l)$.

The polynomials g_j in the theorem can be obtained by replacing x_i with $\frac{y_i}{1+y_i}$ as is well-known.

Theorem 22 (Theorem 21 continued). *Thus the Dixon matrix with respect to Bernstein basis of degree $\overline{N}'$ and N' can be obtained from the Dixon polynomial of F viewed as bilinear form in the Bernstein basis elements $\beta_{\overline{N}',\overline{I}}(\overline{X})$ and $\beta_{N',I}(X)$. Its entry with row index $\beta_{\overline{N}',\overline{I}}(\overline{X})$ and column index $\beta_{N,I}(X)$ can be obtained by multiplying the entry with row index $\overline{Y}^{\overline{I}}$ and column index Y^I of the Dixon matrix of G by $\left(\frac{\overline{N}'}{\overline{I}}\right)^{-1} \cdot \left(\frac{N'}{I}\right)^{-1}$.*

Remark 23. Theorem 21 implies that the Dixon matrix of F with respect to the scaled Bernstein basis is the Dixon matrix of G.

4.2 Mixed Bernstein Basis Degrees

Theorem 21 requires that the Bernstein basis degrees of the f_j's are all the same (unmixed). In this section we treat the case of not-necessarily unmixed Bernstein basis degrees by lifting to unmixed degrees. Let us first consider a simple example.

Example 24. Let

$$f_0 = 3\,\beta_{2,2}(x) + 5\,\beta_{2,1}(x) + 7\,\beta_{2,0}(x),$$
$$f_1 = a_{11}\,\beta_{1,1}(x) + a_{10}\,\beta_{1,0}(x)$$

such that the Bernstein basis degree of f_0 is 2 and of f_1 is 1. In order to be able to apply Theorem 21 one can lift (by a well-known procedure) the basis degree of f_1 to 2. One obtains

$$f_1 = a_{11}\,\beta_{2,2}(x) + \frac{a_{11}+a_{10}}{2}\,\beta_{2,1}(x) + a_{10}\,\beta_{2,0}(x).$$

Hence, according to Theorem 21, $G = (g_0, g_1)$ with

$$g_0 = 3\,x^2 + 5\,x + 7,$$
$$g_1 = a_{11}\,x^2 + (a_{11}+a_{10})\,x + a_{10}.$$

Thus the Dixon polynomial θ_F of $F = (f_0, f_1)$ with respect to the Bernstein basis is given by

$$\theta_F = (1-\overline{x}) \cdot (1-x) \cdot \theta_G\left(\frac{\overline{x}}{1-\overline{x}}, \frac{x}{1-x}\right)$$
$$= (3a_{10} - 7a_{11})\,\beta_{1,1}(\overline{x})\,\beta_{1,1}(x) + (3a_{10} - 7a_{11})\,\beta_{1,1}(\overline{x})\,\beta_{1,0}(x)$$
$$+ (3a_{10} - 7a_{11})\,\beta_{1,0}(\overline{x})\,\beta_{1,1}(x) + (3a_{10} - 7a_{11})\,\beta_{1,0}(\overline{x})\,\beta_{1,0}(x).$$

As in the previous example, given $F = (f_0, \ldots, f_l)$ with mixed basis degrees $N_j = (n_{j1}, \ldots, n_{jl})$ we first lift the basis degrees of the f_j's to $N_{\max} = (\max_j(n_{j1}), \ldots \max_j(n_{jl}))$. Then we apply the construction from Sect. 4.1 to the lifted f_j's.

But, when following this approach the resulting Dixon matrix can become larger than actually needed! This is shown in the following example.

Example 25 (Example 24 continued). The Dixon matrix with respect to Bernstein basis of the lifted f_j's is

$$\begin{pmatrix} 3a_{10} - 7a_{11} \; 3a_{10} - 7a_{11} \\ 3a_{10} - 7a_{11} \; 3a_{10} - 7a_{11} \end{pmatrix}.$$

Observe that the maximal minor $3a_{10} - 7a_{11}$ of the Dixon matrix is the determinant of a 1-by-1 sub-matrix. That is, the size of the Dixon matrix is larger than necessary. This larger size is caused by the lifting operation that lifted the Bernstein basis degree of f_1.

Since the f_j's are lifted, the resulting Dixon polynomial is lifted as well, that is, its Bernstein basis has higher degree then necessary. This causes the enlarged Dixon matrix.

Example 26 (Example 25 continued). The Dixon polynomial θ_F can also be represented as $(3a_{10} - 7a_{11}) \beta_{0,0}(\overline{x}) \beta_{0,0}(x)$ with basis degree 0.

Because of the possibility of representing the Dixon polynomial with lower basis degree, as is shown in the previous example, we describe a procedure to reduce the basis degree of a Bernstein polynomial (Definitions 27 and 28).

Definition 27 ([33]). *Let the head and tail operators be defined as*

$$\mathrm{Head}_m\left(x, \sum_{j=0}^{m} a_j \, \beta_{j,m}(x)\right) = \sum_{j=0}^{m} (-1)^{m-j} \binom{m}{j} a_j$$

and

$$\mathrm{Tail}_m\left(x, \sum_{j=0}^{m} a_j \, \beta_{j,m}(x)\right) =$$

$$\sum_{j=0}^{m-1} (-1)^j \left(\sum_{i=0}^{j} (-1)^i \binom{m}{i} a_i \right) \binom{m-1}{j}^{-1} \beta_{j,m-1}(x).$$

The head operator $\mathrm{Head}_m(x, \, p)$ returns the coefficient of x^m of the polynomial p of Bernstein basis degree m. The tail operator $\mathrm{Head}_m(x, \, p)$ returns p represented with respect to Bernstein basis of degree $m - 1$ if $\mathrm{Head}_m(x, \, p) = 0$.

Definition 28. (Basis degree reduction algorithm)

Input: *Bernstein polynomial p in variables $x_1, \ldots, x_l$*
Output: *Bernstein polynomial q of minimal Bernstein basis degree in each variable x_i with $p = q$*
Algorithm: *For each variable x_i of p, repeat $p \leftarrow \text{Tail}_m(x_i, p)$, where m is the Bernstein basis degree of p in x_i, until $\text{Head}_m(x_i, p) \neq 0$. Then return $q \leftarrow p$.*

The correctness of the algorithm is shown in the appendix (Theorem 37).

Example 29 (Example 26 continued).

$$\theta_F = ((3a_{10} - 7a_{11})\, \beta_{1,1}(x) + (3a_{10} - 7a_{11})\, \beta_{1,0}(x))\, \beta_{1,1}(\overline{x}) +$$
$$((3a_{10} - 7a_{11})\, \beta_{1,1}(x) + (3a_{10} - 7a_{11})\, \beta_{1,0}(x))\, \beta_{1,0}(\overline{x}).$$

Since $\text{Head}_1(\overline{x}, \theta_F) = 0$, we compute

$$p \leftarrow \text{Tail}_1(\overline{x}, \theta_F) = ((3a_{10} - 7a_{11})\, \beta_{1,1}(x) + (3a_{10} - 7a_{11})\, \beta_{1,0}(x))\, \beta_{0,0}(\overline{x}).$$

Since now $\text{Head}_0(\overline{x}, p) \neq 0$ and $\text{Head}_1(x, p) = 0$, we compute

$$p \leftarrow \text{Tail}_1(p) = (3a_{10} - 7a_{11})\, \beta_{0,0}(\overline{x})\, \beta_{0,0}(x)$$

which is the desired Dixon polynomial of minimal Bernstein basis degree $(0, 0)$.

Remark 30. Concluding, the basis degree reduction algorithm allows us to minimize the Bernstein basis degree of the Dixon polynomial for polynomials of mixed basis degrees. Thus the Dixon matrix obtained from the Dixon polynomial after applying the basis degree reduction algorithm is also of minimal size.

4.3   Dixon Resultant

We study the relationship between the Dixon resultant computed with respect to Bernstein and with respect to power basis. Before stating the main theorem, we introduce the natural definitions of gcd of maximal minors of Dixon matrices (rank sub-matrix constructions [26] of Kapur/Saxena/Yang) with respect to the different bases.

Definition 31. *Let $F = (f_0, \ldots, f_l)$ be a list of l-variate polynomials with parametric (polynomial) coefficients. Then*

1. *the power-basis gcd of the maximal Dixon minors and*
2. *the Bernstein-basis gcd of the maximal Dixon minors*

is defined (up to constant factor) as the gcd of all maximal minors of the Dixon matrix of F constructed, respectively, with respect to

1. the power basis as in [26], and
2. the Bernstein basis as it is described in the previous two sub-sections.

Theorem 32. *The Bernstein-basis gcd of the maximal Dixon minors of F equals (up to constant factor) the power-basis gcd of the maximal Dixon minors of F.*

We conclude with a short example for Theorem 32.

Example 33. Consider

$$
\begin{aligned}
f_0 &= a_{11}\, x_1\, x_2 + a_{10}\, x_1 + a_{01}\, x_2 + a_{00}, \\
f_1 &= b_1\, x_1 + b_0, \\
f_2 &= c_1\, x_1 + c_0.
\end{aligned}
$$

Notice that f_1 and f_2 are both univariate in the variable x_1, whereas f_0 is bivariate. Therefore the resultant of this system is the resultant of f_1 and f_2 which is $r = b_1\, c_0 - b_0\, c_1$. The 2-by-1 Dixon matrix with respect to the power basis is

$$
\begin{pmatrix}
-a_{01}\, (b_1\, c_0 - b_0\, c_1) \\
-a_{11}\, (b_1\, c_0 - b_0\, c_1)
\end{pmatrix}.
$$

The gcd of the maximal minors is r. In Bernstein basis representation we have

$$
\begin{aligned}
f_0 &= (a_{00} + a_{01} + a_{10} + a_{11})\, \beta_{N,(1,1)}(X) + (a_{00} + a_{10})\, \beta_{N,(1,0)}(X) \\
&\quad + (a_{00} + a_{01})\, \beta_{N,(0,1)}(X) + a_{00}\, \beta_{N,(0,0)}(X), \\
f_1 &= (b_0 + b_1)\, \beta_{N,(1,1)}(X) + (b_0 + b_1)\, \beta_{N,(1,0)}(X) + b_0\, \beta_{N,(0,1)}(X) \\
&\quad + b_0\, \beta_{N,(0,0)}(X), \\
f_2 &= (c_0 + c_1)\, \beta_{N,(1,1)}(X) + (c_0 + c_1)\, \beta_{N,(1,0)}(X) + c_0\, \beta_{N,(0,1)}(X) \\
&\quad + c_0\, \beta_{N,(0,0)}(X),
\end{aligned}
$$

where $N = (1,1)$ and $X = (x_1, x_2)$. The minimal size Dixon matrix with respect to Bernstein basis is

$$
\begin{pmatrix}
-(a_{11} + a_{01})\, (b_1\, c_0 - b_0\, c_1) \\
-a_{01}\, (b_1\, c_0 - b_0\, c_1)
\end{pmatrix}.
$$

The gcd of the maximal minors is also r as it is stated in Theorem 32.

5   Applications

In Sect. 5.1 and 5.2 we respectively consider a proof of a geometric theorem and surface-curve intersections in Bernstein basis. Furthermore, in Sect. 5.3 we investigate the efficiency of Bernstein-basis resultants for several well-known geometric benchmark problems, which also include some of the systems from Sect. 5.2 and 5.1

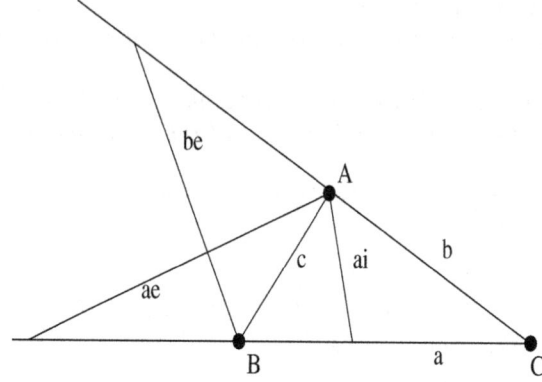

Fig. 1. Bisectors

5.1   A Theorem on Drawing by Compass and Ruler

As in [14,24] we consider the question whether a triangle can by drawn by compass and ruler given the lengths of certain external and internal bisectors. The triangle ABC can be seen in Fig. 1. The bisectors are denoted by ai, ae, and be. As shown in [14], the squares of the length of the bisectors depend rationally on the lengths a, b, c of the sides of the triangle. Thus one obtains a corresponding system of three polynomials in the lengths of the bisectors and of the sides [14]. When converted into Bernstein basis, two polynomials in this system have a square Newton polygon with respect to the Bernstein basis, namely, the convex hull of the points $\{(0,0), (3,0), (0,3), (3,3)\}$. The Newton polygon of the third polynomial is spanned by the points $\{(2,0), (0,2), (3,0), (0,3)\}$. The coefficients of the polynomials are quite large and therefore the system is not included here. Still the Dixon resultant, eliminating b and c, of the polynomial system can be computed quite efficiently because the corresponding Dixon matrix is only of size 18×18 and of rank 9. Since the resultant of the polynomial system is of degree 20, it can be shown that [14] it is impossible to construct the triangle ABC with compass and ruler from its bisectors.

5.2   Surface-Curve Intersection

We use Bernstein-basis Dixon resultants in order to intersect a real surface

$$(x, y, z) = (s_1(r, s), s_2(r, s), s_3(r, s))$$

with real curves $(c_1(t), c_2(t), c_3(t))$. For example, consider the Enneper surface [14] in Fig. 2. Enneper surface is named after the German mathematician Alfred Enneper who constructed the surface in 1863. This is a well known minimal surface, that is, a surface with vanishing mean curvature. Its Bernstein-basis

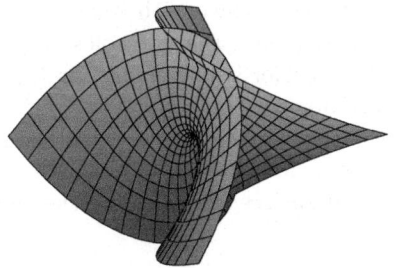

Fig. 2. Enneper surface

representation is

$$s_1 = 5/3\,\beta_{N_1,(3,2)}(r,s) + 2/3\,\beta_{N_1,(3,1)}(r,s) + 2/3\,\beta_{N_1,(3,0)}(r,s)$$
$$+ 4/3\,\beta_{N_1,(2,2)}(r,s) + 2/3\,\beta_{N_1,(2,1)}(r,s) + 2/3\,\beta_{N_1,(2,0)}(r,s)$$
$$+ 2/3\,\beta_{N_1,(1,2)}(r,s) + 1/3\,\beta_{N_1,(1,1)}(r,s) + 1/3\,\beta_{N_1,(1,0)}(r,s)$$
$$s_2 = -5/3\,\beta_{N_2,(2,3)}(r,s) - 4/3\,\beta_{N_2,(2,2)}(r,s) - 2/3\,\beta_{N_2,(2,1)}(r,s)$$
$$- 2/3 - 2/3\,\beta_{N_2,(1,3)}(r,s)\,\beta_{N_2,(1,2)}(r,s) - 1/3\,\beta_{N_2,(1,1)}(r,s)$$
$$- 2/3\,\beta_{N_2,(0,3)}(r,s) - 2/3\,\beta_{N_2,(0,2)}(r,s) - 1/3\,\beta_{N_2,(0,1)}(r,s)$$
$$s_3 = \beta_{N_3,(2,1)}(r,s) + \beta_{N_3,(2,0)}(r,s) - \beta_{N_3,(2,2)}(r,s) - \beta_{N_3,(0,2)}(r,s),$$

where $N_1 = (3,2)$, $N_2 = (2,3)$, $N_3 = (2,2)$ are the Bernstein basis degrees of the corresponding polynomials s_1, s_2, s_3. Furthermore, consider the curve

$$c_1 = 11/3\,\beta_{2,2}(t) + \beta_{2,1}(t) + \beta_{2,0}(t),$$
$$c_2 = -3\,\beta_{2,2}(t) + \beta_{2,1}(t) + \beta_{2,0}(t),$$
$$c_3 = -29/3\,\beta_{2,2}(t) + \beta_{2,1}(t) + \beta_{2,0}(t).$$

Next, we construct the differences $s_1 - x$, $s_2 - y$, $s_3 - z$, where x, y, z are some independent symbols. In order to obtain these differences in proper Bernstein-basis representation, one views the symbols x, y, z as constant polynomials in r and s. Thus they respectively correspond to $x \cdot \sum_{i=0}^{3} \sum_{j=0}^{2} \beta_{N_1,(i,j)}(r,s)$, $y \cdot \sum_{i=0}^{2} \sum_{j=0}^{3} \beta_{N_2,(i,j)}(r,s)$, and $z \cdot \sum_{i=0}^{2} \sum_{j=0}^{2} \beta_{N_3,(i,j)}(r,s)$. Therefore we get

$$f_1 = s_1 - x \cdot \sum_{i=0}^{3} \sum_{j=0}^{2} \beta_{N_1,(i,j)}(r,s),$$

$$f_2 = s_2 - y \cdot \sum_{i=0}^{2} \sum_{j=0}^{3} \beta_{N_2,(i,j)}(r,s),$$

$$f_3 = s_3 - z \cdot \sum_{i=0}^{2} \sum_{j=0}^{2} \beta_{N_3,(i,j)}(r,s).$$

In order to intersect the Enneper surface with the curve, we substitute (c_1, c_2, c_3) for (x, y, z) in the f_i's and eliminate the variables r, s. However, it depends on the desired application if we prefer substituting for (x, y, z) before eliminating, or eliminating before substituting. The latter approach may be more efficient if we expect to intersect several different curves with the surface. For the current example, we take this approach. Thus, eliminating r and s from f_1, f_2, f_3 by using the Bernstein-basis Dixon resultant yields the implicit representation of the Enneper surface

$$
\begin{aligned}
& 46656\, x^6 - 46656\, x^4 z + 77760\, z^3 x^4 + 279936\, x^4 z^2 - 139968\, x^4 y^2 \\
& + 248832\, z^5 x^2 - 248832\, z^3 x^2 + 404352\, z^3 x^2 y^2 + 27648\, z^6 x^2 + 139968\, y^4 x^2 \\
& + 93312\, z y^2 x^2 + 414720\, z^4 x^2 + 73728\, z^7 - 331776\, z^5 - 248832\, y^2 z^3 - 414720\, z^4 y^2 \\
& - 279936\, y^4 z^2 - 46656\, y^4 z - 46656\, y^6 - 4096\, z^9 + 248832\, z^5 y^2 - 27648\, z^6 y^2 \\
& \hspace{9cm} + 77760\, y^4 z^3.
\end{aligned}
$$

Then, substituting (c_1, c_2, c_3) for (x, y, z) yields

$$
\begin{aligned}
p = {} & -87008774400\, \beta_{18,18}(t) - 12938372352\, \beta_{18,17}(t) - \frac{942126336}{17}\, \beta_{18,16}(t) \\
& + \frac{11214109440}{17}\, \beta_{18,15}(t) + \frac{13965831936}{85}\, \beta_{18,14}(t) - \frac{3177969408}{119}\, \beta_{18,13}(t) \\
& - \frac{49317661440}{1547}\, \beta_{18,12}(t) - \frac{1563215104}{221}\, \beta_{18,11}(t) + \frac{8489107712}{2431}\, \beta_{18,10}(t) \\
& + \frac{2213933312}{715}\, \beta_{18,9}(t) + \frac{1585319168}{2431}\, \beta_{18,8}(t) - \frac{4070656}{17}\, \beta_{18,7}(t) \\
& + \frac{1526528}{17}\, \beta_{18,6}(t) + \frac{9185024}{17}\, \beta_{18,5}(t) + \frac{55624448}{85}\, \beta_{18,4}(t) \\
& + \frac{8990464}{17}\, \beta_{18,3}(t) + \frac{6371072}{17}\, \beta_{18,2}(t) + 297728\, \beta_{18,1}(t) + 297728\, \beta_{18,0}(t).
\end{aligned}
$$

Now, one can approximate the real roots of p using standard software. Here, the example has been chosen such that one obtains $t = \pm\frac{1}{2}$. All the other roots of p are complex and thus do not correspond to the intersection of the real Enneper surface with the real curve.

5.3 Geometric Benchmarks

We study several geometric benchmark problems from [14]:

- bisector (Sect. 11.1.8),
- strophoid (Sect. 11.2.1),
- surface (Sect. 11.2.2),
- sphere (Sect. 11.2.3),
- bicubic (Sect. 11.2.4),
- cubic (Sect. 11.2.5),
- enneper (Sect. 11.2.6).

| Benchmark | Bernstein basis | | | | Power basis | | | |
| | Dixon matrix | | Macaulay matrix | | Dixon matrix | | Macaulay matrix | |
	Size	Rank	Size	Rank	Size	Rank	Size	Rank
bisector	18×18	9	136×136	101	14×13	13	55×55	45
strophoid	8×8	4	36×36	30	6×5	5	36×36	30
surface	18×18	5	153×153	110	12×12	5	36×36	30
sphere	16×16	10	91×91	91	11×12	10	45×45	41
bicubic	18×18	18	153×153	123	18×18	18	66×66	66
cubic	18×18	11	105×105	87	11×11	11	36×36	34
enneper	18×18	9	91×91	70	11×11	9	28×28	28

Fig. 3. Resultant matrices for Bernstein and power bases

For more details on the benchmark "bisector" which represents a theorem from elementary geometry see Sect. 5.1. The other benchmarks are surface implicitization problems. The implicitization corresponding to the benchmark "enneper" is detailed in Sect. 5.2. The following table shows the dimensions and ranks of the Dixon and Macaulay matrices obtained for the benchmark problems. The data with respect to power basis is from [14] whereas the data with respect to Bernstein basis has been generated by the authors using the Maple packages [32,34] after converting the polynomials from [14] into Bernstein basis.

We observe in Fig. 3 that the ranks of the Bernstein-basis Dixon matrices are less than or equal to the power-basis Dixon matrices. This means that extracting resultants from maximal-rank minors can be performed at least as efficiently in Bernstein basis as in power basis. The sizes of the Bernstein-basis Dixon matrices tend to be slightly larger than the corresponding power-basis Dixon matrices. However, this only insignificantly affects running times. That is, current efficient methods for extracting maximal-rank minors of matrices first find a submatrix of maximal rank and then compute its determinant (see e.g. [8]). Since computing the determinant is by far the more expensive operation, the practical running time most significantly depends on the rank of the matrix, that is, of the size of its maximal-rank submatrix rather than the size of the whole matrix.

In contrast, we see in Fig. 3 that the ranks and sizes of the Bernstein-basis Macaulay matrices are greater than or equal to the corresponding ranks of the power-basis Macaulay matrices for the benchmarks. Moreover, we observed that the corresponding Macaulay resultants vanish and therefore do not yield any information about the existence of common roots, for both Bernstein and power basis representations. However, we noticed that for these benchmarks the maximal-rank minors of the Macaulay matrix still are projection operators. Similar observations have also been made by [27].

Furthermore, we point out that [14] contains several other polynomial systems from applications in elementary geometry not shown in Fig. 3. Those systems are sparse in power basis representation. But they become dense and suffer from a huge blow up of coefficient size when converted into Bernstein basis. We found that for those systems the Bernstein basis representation seems inefficient and causes the construction of Dixon matrices and the extraction of maximal-rank minors to take many hours.

6   Conclusion and Future Work

We have adapted the construction of multivariate resultant formulations to be applicable to parametrized multivariate polynomial systems in Bernstein basis. The key idea is to transform a polynomial system represented in Bernstein basis by replacing each variable x_i by $\frac{y_i}{1+y_i}$, where y_i is a new variable, and relating the zero set of the original polynomial system to that of the transformed polynomial system. This transformation on a polynomial system in Bernstein basis leads to a new polynomial system in the standard power basis preserving the coefficients of the original system. So the classical Macaulay construction applies leading to a resultant matrix with respect to the Bernstein basis. The main result about Macaulay resultant formulation is that the total degree resultant of the polynomial system in Bernstein basis vanishes if and only if the polynomial system has a common Bernstein-toric root, a common infinite root, or the leading forms of the transformed polynomial system have a common non-trivial root.

In the case of the Dixon resultant formulation the rank sub-matrix constructions for the original system and the transformed system are essentially equivalent. Thus, known results about exactness of Dixon resultants of a sub-class of polynomial systems as discussed in [12] carry over to polynomial systems represented in the Bernstein basis. Furthermore, in certain cases, when the extraneous factor in a projection operator constructed from the Dixon resultant formulation is precisely known, the projection operator of a polynomial system in the Bernstein basis can be predicted a priori to precisely include the known extraneous factor.

We conjecture that all the known results about the exactness of resultant matrices vis a vis resultant computation and the nature of extraneous factors in projection operators when polynomial systems are not generic and/or specialized extend to the case when a polynomial system is represented in Bernstein basis. This needs to be explored further.

Applications of these results have been discussed in the context of geometry theorem proving, implicitization and intersection of surfaces with curves. It was observed that for geometry problems, where power basis representation of polynomials is sparse, their representation in Bernstein bases often becomes dense; as a result, the Bernstein basis representation is inefficient.

In future work, we propose to study how various resultant matrix constructions behave when the coefficients of polynomials in Bernstein basis are floating points, instead of parameters. We would like to understand the numerical stability of the Dixon matrix construction and how well-behaved the resultant matrices are, e.g. by studying their condition numbers [40].

References

1. Amiraslani, A.: Dividing polynomials when you only know their values. In: Gonzalez-Vega, L., Recio, T. (eds.) Proceedings of Encuentros de Álgebra Computacional y Aplicaciones (EACA) 2004, pp. 5–10 (2004)
2. Barnett, S.: Polynomials and linear control systems. Monographs and Textbooks in Pure and Applied Mathematics, vol. 77. Marcel Dekker Inc., New York (1983)
3. Barnett, S.: Division of generalized polynomials using the comrade matrix. Linear Algebra Appl. 60, 159–175 (1984)
4. Barnett, S.: Euclidean remainders for generalized polynomials. Linear Algebra Appl. 99, 111–122 (1988)
5. Berchtold, J., Bowyer, A.: Robust arithmetic for multivariate bernstein-form polynomials. In: Computer-Aided Design, pp. 681–689 (2000)
6. Bini, D.A., Gemignani, L.: Bernstein-Bezoutian matrices. Theoret. Comput. Sci. 315(2-3), 319–333 (2004)
7. Bini, D.A., Gemignani, L., Winkler, J.R.: Structured matrix methods for CAGD: an application to computing the resultant of polynomials in the Bernstein basis. Numer. Linear Algebra Appl. 12(8), 685–698 (2005)
8. Brazier, M., Chcherba, A.: MatDetInterp. Symbolic matrix determinant interpolator, http://www.chtcherba.com/arthur/Projects/MatDetInterp/
9. Busé, L., Elkadi, M., Mourrain, B.: Generalized resultants over unirational algebraic varieties. J. Symbolic Computation 29(4-5), 515–526 (2000)
10. Canny, J.: Generalised characteristic polynomials. J. Symbolic Computation 9, 241–250 (1990)
11. Cheng, H., Labahn, G.: On computing polynomial GCDs in alternate bases. In: ISSAC 2006, pp. 47–54. ACM, New York (2006)
12. Chtcherba, A., Kapur, D.: Exact resultants for corner-cut unmixed multivariate polynomial systems using the Dixon formulation. J. Symbolic Computation 36(3-4), 289–315 (2003)
13. Chtcherba, A.D., Kapur, D., Minimair, M.: Cayley-dixon resultant matrices of multi-univariate composed polynomials. In: Ganzha, V.G., Mayr, E.W., Vorozhtsov, E.V. (eds.) CASC 2005. LNCS, vol. 3718, pp. 125–137. Springer, Heidelberg (2005)
14. Chtcherba, A.D.: A new Sylvester-type Resultant Method based on the Dixon-Bézout Formulation. PhD dissertation, University of New Mexico, Department of Computer Science (August 2003)
15. Chtcherba, A.D., Kapur, D.: Conditions for determinantal formula for resultant of a polynomial system. In: ISSAC 2006: Proceedings of the 2006 International Symposium on Symbolic and Algebraic Computation, Genoa, Italy, pp. 55–62. ACM, New York (2006), doi:10.1145/1145768.1145784
16. Corless, R.: Generalized companion matrices in the lagrange basis. In: Gonzalez-Vega, L., Recio, T. (eds.) Proceedings of Encuentros de Álgebra Computacional y Aplicaciones (EACA) 2004, pp. 317–322 (2004)
17. Cox, D., Little, J., O'Shea, D.: Using Algebraic Geometry. Springer, Heidelberg (1998)
18. D'Andrea, C.: Macaulay style formulas for sparse resultants. Trans. Amer. Math. Soc. 354(7), 2595–2629 (electronic) (2002)
19. Devore, R.A., Lorentz, G.G.: Constructive Approximation. Springer, Heidelberg (1993)

20. Diaz-Toca, G.M., Gonzalez-Vega, L.: Barnett's theorems about the greatest common divisor of several univariate polynomials through Bezout-like matrices. J. Symbolic Comput. 34(1), 59–81 (2002)
21. Dixon, A.-L.: On a form of the elimination of two quantics. Proc. London Math. Soc. 6, 468–478 (1908)
22. Farin, G.F.: Curves and Surfaces for CAGD: A practical guide, 5th edn. Morgan Kaufmann, San Francisco (1991)
23. Gemignani, L.: Manipulating polynomials in generalized form. Tech. Rep. TR-96-14, Università di Pisa, Departmento di Informatica, Corso Italia 40, 56125 Pisa, Italy (December 1996)
24. Heymann, W.: Problem der Winkelhalbierenden. Ztschr. f. Math. und Phys. 35 (1890)
25. Kapur, D., Saxena, T.: Sparsity considerations in Dixon resultants. In: Proceedings of the Twenty-Eighth Annual ACM Symposium on the Theory of Computing, Philadelphia, PA, pp. 184–191. ACM, New York (1996)
26. Kapur, D., Saxena, T., Yang, L.: Algebraic and geometric reasoning using the Dixon resultants. In: ACM ISSAC 1994, Oxford, England, pp. 99–107 (July 1994)
27. Lewis, R.: Comparing acceleration techniques for the Dixon and Macaulay resultants. Mathematics and Computers in Simulation (2008) (accepted)
28. Macaulay, F.S.: The algebraic theory of modular systems. Cambridge Mathematical Library (1916)
29. Mani, V., Hartwig, R.E.: Generalized polynomial bases and the Bezoutian. Linear Algebra Appl. 251, 293–320 (1997)
30. Manocha, D., Krishnan, S.: Algebraic pruning: A fast technique for curve and surface intersection. Computer-Aided Geometric Design 20, 1–23 (1997)
31. Maroulas, J., Barnett, S.: Greatest common divisor of generalized polynomial and polynomial matrices. Linear Algebra Appl. 22, 195–210 (1978)
32. Minimair, M.: MR, macaulay resultant package for Maple (April 2003), http://minimair.org/MR.mpl
33. Minimair, M.: Basis-independent polynomial division algorithm applied to division in lagrange and bernstein basis (CD-ROM). In: Kapur, D. (ed.) Proceedings of Asian Symposium on Computer Mathematics (ASCM). National University of Singapore (2007)
34. Minimair, M.: DR, Maple package for computing Dixon projection operators (resultants) (2007), http://minimair.org/dr
35. Tsai, Y.-F., Farouki, R.T.: Algorithm 812: BPOLY: An object-oriented library of numerical algorithms for polynomials in Bernstein form. ACM Transactions on Mathematical Software 27(2), 267–296 (2001)
36. Winkler, J.R.: A resultant matrix for scaled Bernstein polynomials. Linear Algebra Appl. 319(1-3), 179–191 (2000)
37. Winkler, J.R.: Computational experiments with resultants for scaled Bernstein polynomials. In: Mathematical Methods for Curves and Surfaces, Oslo, pp. 535–544. Innov. Appl. Math., Vanderbilt Univ. Press, Nashville, TN (2001)
38. Winkler, J.R.: Properties of the companion matrix resultant for Bernstein polynomials. In: Uncertainty in Geometric Computations. Kluwer Internat. Ser. Engrg. Comput. Sci., vol. 704, pp. 185–198. Kluwer Acad. Publ., Boston (2002)
39. Winkler, J.R.: A companion matrix resultant for Bernstein polynomials. Linear Algebra Appl. 362, 153–175 (2003)
40. Winkler, J.R.: Numerical and algebraic properties of Bernstein basis resultant matrices. In: Computational Methods for Algebraic Spline Surfaces, pp. 107–118. Springer, Berlin (2005)

41. Winkler, J.R., Goldman, R.N.: The Sylvester resultant matrix for Bernstein polynomials. In: Curve and Surface Design, Saint-Malo. Mod. Methods Math., pp. 407–416. Nashboro Press, Brentwood (2003)

A   Appendix: Proofs

A.1   Proofs for Macaulay-Style Construction

Proof (Theorem 8). Consider the polynomial

$$f = \sum_{\substack{I \leq N, \\ |I| \leq m}} a_I \, \beta_{N,I}(X)$$

with total-degree Bernstein basis support where $N = (n_1, \dots, n_l)$. Since by definition $1 \leq m \leq n_j$, we have $|N| \geq l \cdot m > m$ (compare also Remark 4). Therefore f does not contain the basis element $\beta_{N,N}(X) = x_1^{n_1} \cdots x_n^{n_l}$ which does not vanish for $x_j = 1$. Moreover, all the basis elements $\beta_{N,I}(X)$ with some $i_j < n_j$ contain the factor $(1 - x_j)^{n_j - i_j}$ which vanishes for $x_j = 1$.

For proving Theorem 16, we start with a simple lemma about the range of the variable transformation used to map f_j into g_j. It is well-known that f_j can be transformed into g_j by substituting $\frac{y_i}{1+y_i}$ into x_i and by multiplying the result by $(1 + y_i)^{N_{ji}}$ for all i.

Lemma 34. *Let K be a field. Then the map $y \mapsto x = \frac{y}{1+y}$ bijectively maps $K\backslash\{-1\}$ to $K\backslash\{1\}$ with inverse $x \mapsto \frac{x}{1-x}$.*

Proof (Lemma 34). Obviously the mapping is only defined for $K\backslash\{-1\}$. It remains to show that there is no y such that $x = 1$. This is true because the equation $\frac{y}{1+y} = 1$ does not have a solution. Bijectivity and inverse follow from solving $\frac{y}{1+y} = x$.

The next lemma shows an alternative way of constructing g_j from f_j by homogenization.

Lemma 35. *g_j of Theorem 16 can be obtained from f_j by homogenizing g_j with respect to each variable x_i with, say, homogenizing variable z_i and by replacing (x_i, z_i) with $(y_i, 1 + y_i)$.*

Proof (Lemma 35). Since the transformation from g_j into f_j is composed of individual independent transformations for each variable x_i, it is sufficient to show the lemma for transforming the Bernstein basis element $f = \beta_{n,k}(x)$ into $g = \binom{n}{k} y^k$. Now,

$$g = \binom{n}{k} y^k \left((1+y) - y\right)^{n-k}$$

$$= \binom{n}{k} x^k \, (z - x)^{n-k},$$

where $x = y$ and $z = 1 + y$ and $\binom{n}{k} x^k \, (z - x)^{n-k}$ is the homogenization of f.

Now we are ready to prove the theorem.

Proof (Theorem 16). As is well-known, g_j can be obtained from f_j by substituting $\frac{y_i}{1+y_i}$ for x_i and by multiplying the result by $(1 + y_i)^{N_{ji}}$. We will use this fact several times below without explicitly referring to it.

Now, let us study when the Macaulay (projective) resultant of the g_j's vanishes. It vanishes if and only if

1. either the g_j's have a common root or the
2. leading forms (1)

have a common non-trivial root.

We analyze Case 1. Let Y be a common root of the g_j's. First we assume that $y_i \neq -1$ for all i. By Lemma 34, we have that $x_i \neq 1$ for all i. Therefore such a Y yields a Bernstein-toric root X of the f_j's. Next we assume that there is an i such that $y_i = -1$. By Lemma 34 there is no corresponding tuple X. However, by Lemma 35 a root with $y_i = -1$ can be interpreted as a root of the leading coefficients, with respect to x_i, of the f_j's as it is the usual practice of algebraic geometry.

A.2   Proofs for Bézout/Cayley/Dixon-Style Construction

First we formulate a proposition to be used in the proof of Theorem 21.

Proposition 36. *Let* $\overline{y} = \frac{\overline{x}}{1-\overline{x}}$ *and* $y = \frac{x}{1-x}$. *Then*

$$x - \overline{x} = (y - \overline{y}) (1 - \overline{x}) (1 - x)$$

Proof (Proposition 36). We have

$$(y - \overline{y}) (1 - \overline{x}) (1 - x) = \left(\frac{x}{1 - x} - \frac{\overline{x}}{1 - \overline{x}} \right) (1 - \overline{x}) (1 - x)$$

$$= x \, (1 - \overline{x}) - \overline{x} \, (1 - x)$$

$$= x - \overline{x}.$$

Now we are ready to prove the theorem.

Proof (Theorem 21). By Proposition 36 and with $\overline{y}_j$, y_j respectively denoting $\frac{\overline{x}_j}{1-\overline{x}_j}$ and $\frac{x_j}{1-x_j}$, the Dixon polynomial of F is

$$\theta_F = \begin{vmatrix} \cdots & f_j(x_1, \ldots, x_l) & \cdots \\ \cdots & f_j(\overline{x}_1, \ldots, x_l) & \cdots \\ & \vdots & \\ \cdots & f_j(\overline{x}_1, \ldots, \overline{x}_l) & \cdots \end{vmatrix} \bigg/ \prod_{i=1}^{l} (x_i - \overline{x}_i)$$

$$
\begin{aligned}
&= \frac{\begin{vmatrix} \cdots & (1-x_1)^{n_1} \cdot (1-x_2)^{n_2} \cdots (1-x_l)^{n_l} \cdot g_j(y_1,\ldots,y_l) & \cdots \\ \cdots & (1-\overline{x}_1)^{n_1} \cdot (1-x_2)^{n_2} \cdots (1-x_l)^{n_l} \cdot g_j(\overline{y}_1,\ldots,y_l) & \cdots \\ & \vdots & \\ \cdots & (1-\overline{x}_1)^{n_1} \cdot (1-\overline{x}_2)^{n_2} \cdots (1-\overline{x}_l)^{n_l} \cdot g_j(\overline{y}_1,\ldots,ybar_l) & \cdots \end{vmatrix}}{\prod_{i=1}^{l}(y_i - \overline{y}_i)(1-\overline{x}_i)(1-x_i)}
\end{aligned}
$$

$$
= \begin{vmatrix} \cdots & g_j(y_1,\ldots,y_l) & \cdots \\ \cdots & g_j(\overline{y}_1,\ldots,y_l) & \cdots \\ & \vdots & \\ \cdots & g_j(\overline{y}_1,\ldots,\overline{y}_l) & \cdots \end{vmatrix} \frac{\prod_{i=1}^{l}(1-\overline{x}_i)^{n_i\,(l-i)}(1-x_i)^{n_i\,i}}{\prod_{i=1}^{l}(y_i - \overline{y}_i)(1-\overline{x}_i)(1-x_i)}
$$

$$
= \prod_{i=1}^{l}(1-\overline{x}_i)^{n_i(l-i+1)-1}(1-x_i)^{n_i i-1}\,\theta_G.
$$

The above equality writes θ_F as a linear combination of power products of $\overline{x}$, $1-\overline{x}$, x, $1-x$, that is, of the scaled Bernstein bases in variables $\overline{X}$ and X of degrees $\overline{N}'$ and N'. The linear factors are the coefficients of θ_G. By adjusting the coefficients of θ_G by suitable binomial factors $\left(\overline{\frac{N'}{I}}\right)^{-1}$ and $\left(\frac{N'}{I}\right)^{-1}$ we obtain the coefficients with respect to the Bernstein basis.

Proof (Theorem 22). Theorem 22 follows immediately from the proof of Theorem 21 because the Dixon matrix is the coefficient matrix of the Dixon polynomial.

Theorem 37. *The Bernstein basis degree reduction algorithm of Definition 28 is correct.*

Proof (Theorem 37). Since the algorithm reduces the basis degree independently for each individual variable, it is sufficient to show the correctness for the univariate case $l = 1$. For this case, the correctness follows from repeated application of Theorem 33 of [33] which implies that $p = \mathrm{Head}_m(x_1, p) \cdot x_1^m + \mathrm{Tail}_m(x_1, p)$, where the representation of $\mathrm{Tail}_m(x_1, p)$ has Bernstein basis degree $m - 1$.

Proof (Theorem 32). Let P and B be the Dixon matrix of F with respect to, respectively, the power and the Bernstein basis. Observe that, in the case of unmixed Bernstein basis degree as well as in the case of mixed ones after minimizing the basis degree of the Dixon polynomial (with Definition 28), $P = \overline{T} \cdot B \cdot T$, where $\overline{T}$ and T are basis transformation matrices (for converting the power basis into the Bernstein basis, respectively, for the row and for the column monomial indices of the Dixon matrix P). By Proposition 2.1 of [13] and its preceding remark, the gcd of the maximal minors of B divides the gcd of the maximal minors of P. By inverting $\overline{T}$ and T and reversing the roles of P and B, the gcd of the maximal minors of P divides the gcd of the maximal minors of B and the theorem follows.

Unique Factorization Domains
in the Java Computer Algebra System

Heinz Kredel

IT-Center, University of Mannheim, Germany
kredel@rz.uni-mannheim.de

Abstract. This paper describes the implementation of recursive algorithms in unique factorization domains, namely multivariate polynomial greatest common divisors (gcd) and factorization into irreducible parts in the Java computer algebra library (JAS). The implementation of gcds, resultants and factorization is part of the essential building blocks for any computation in algebraic geometry, in particular in automated deduction in geometry. There are various implementations of these algorithms in procedural programming languages. Our aim is an implementation in a modern object oriented programming language with generic data types, as it is provided by Java programming language. We exemplify that the type design and implementation of JAS is suitable for the implementation of several greatest common divisor algorithms and factorization of multivariate polynomials. Due to the design we can employ this package in very general settings not commonly seen in other computer algebra systems. As for example, in the coefficient arithmetic for advanced Gröbner basis computations like in polynomial rings over rational function fields or (finite, commutative) regular rings. The new package provides factory methods for the selection of one of the several implementations for non experts. Further we introduce a parallel proxy for gcd implementations which runs different implementations concurrently.

Keywords: unique factorization domain, multivariate polynomials, real roots, greatest common divisors.

1 Introduction

We have presented an object oriented design of a Java Computer Algebra System (called JAS in the following) as type safe and thread safe approach to computer algebra in [29–31, 33]. JAS provides a well designed software library using generic types for algebraic computations implemented in the Java programming language. The library can be used as any other Java software package or it can be used interactively or interpreted through an jython (Java Python) front end. The focus of JAS is at the moment on commutative and solvable polynomials, Gröbner bases and applications. By the use of Java as implementation language, JAS is 64-bit and multi-core CPU ready. JAS is available in [37].

This work is interesting for automated deduction in geometry as part of computer algebra and computer science, since it explores the Java [51] type system

T. Sturm and C. Zengler (Eds.): ADG 2008, LNAI 6301, pp. 86–115, 2011.

for expressiveness and eventual short comings. Moreover it employs many Java packages, and stresses their design and performance in the context of computer algebra, in competition with sophisticated computer algebra systems, implemented in other programming languages.

JAS contains interfaces and classes for basic arithmetic of integers, rational numbers and multivariate polynomials with integer or rational number coefficients. The package `edu.jas.gb` contains classes for polynomial and solvable polynomial reduction, Gröbner bases and ideal arithmetic as well as thread parallel and distributed versions of Buchbergers algorithm [35, 36]. Package `edu.-jas.gbmod` contains classes for module Gröbner bases, syzygies for polynomials and solvable polynomials. Applications of Gröbner bases, such as ideal intersections, ideal quotients and Comprehensive Gröbner bases [34] are contained in package `edu.jas.application` and univariate power-series are provided by package `edu.jas.ps`.

In this paper we describe an extension of the library by a package for multivariate polynomial greatest common divisor and factorization computations.

1.1 Related Work

In this section we briefly summarize the discussion of related work from [29]. For an evaluation of the JAS library in comparison to other systems see [30, 31, 33]. For an overview on other computer algebra systems see [20]. The mathematical background for this paper can be found in [28, 17, 10], these books also contain references to the articles where the algorithms have been published first.

Typed computer algebra systems with own programming languages are described in [24, 7, 52]. Computer algebra systems implemented in other programming languages and libraries are: in C/C++ [21, 8, 45], in Modula-2 [38] and in Oberon [22]. Python wrappers for computer algebra systems are presented in [50, 26].

Java computer algebra implementations have been discussed in [55, 43, 44, 4, 13, 2]. Newer approaches are discussed in [46, 25, 14].

The expression of mathematical requirements for generic algorithms in programming language constructs have been discussed in [42, 49]. Object oriented programming concepts in geometric deduction are presented in [40, 9, 56, 23].

This paper contains revised parts of [32]. It is extended by a section on performance comparison with Maple and a description of a class for greatest common divisor computation using (univariate) Hensel lifting. There is a new section on polynomial factorization (Sect. 7) and one on real root isolation (Sect. 8).

1.2 Outline

In the next Sect. 2, we give an overview of the JAS type system for polynomials and an example on using the JAS library. Due to limited space we must assume that you are familiar with the Java programming language [1] and object oriented programming. The introduction is continued by an example of Gröbner base computations over regular rings. The setup and layout of the proposed library extensions for gcd computations are discussed in Sect. 3. The presentation of the main implementing classes in Sect. 4. For the mathematical details see [10, 17, 28]. Sect. 5 presents performance comparisons and Sect. 6 evaluates the presented design. The classes for polynomial factorization are sketched in Sect. 7. A package for real root isolation and real algebraic numbers is discussed in Sect. 8. Finally Sect. 9 draws some conclusions.

2   Introduction to JAS

In this section we introduce the general layout of the polynomial types and show an example for the usage of the JAS library.

Figure 1 shows the central part of the JAS type system. The interface RingElem defines the methods which we expect to be available on all ring

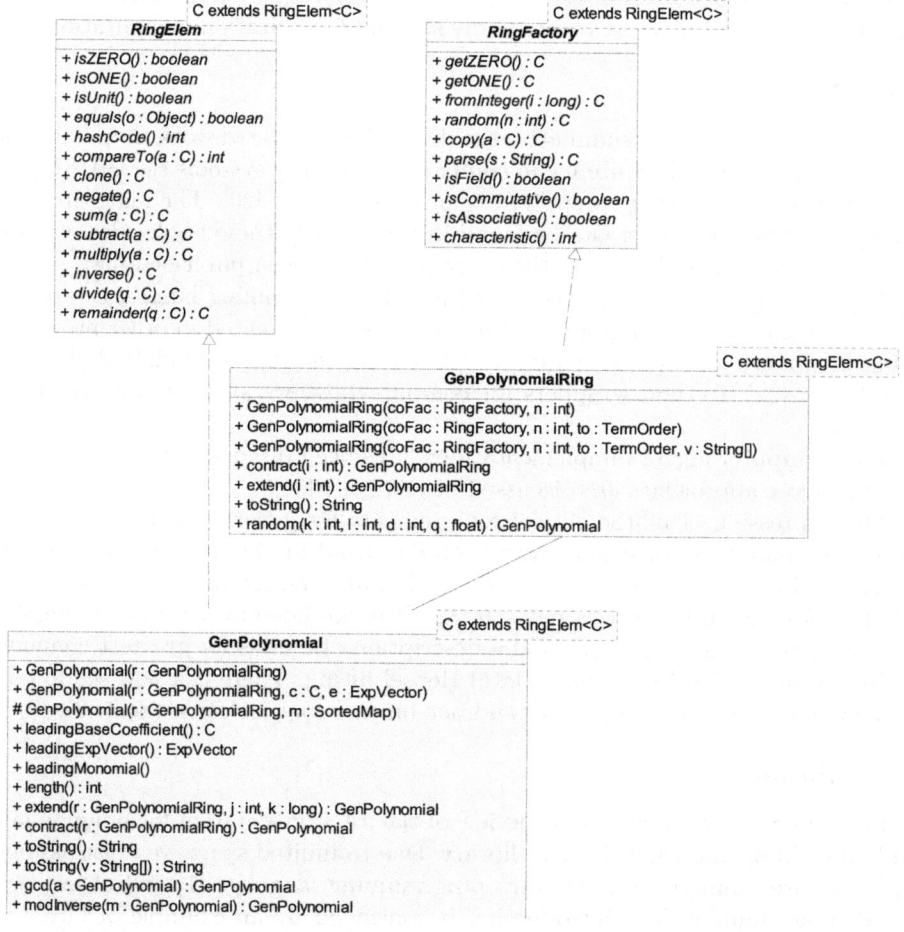

Fig. 1. Overview of the ring element type and of generic polynomials

elements, for example `subtract()`, `multiply()`, `isZERO()` or `isUnit()` with the obvious meanings.

The construction of ring elements is done by factories, modeled after the *factory* creational design pattern [16]. The interface `RingFactory` defines the construction methods, for example `getONE()` to create the one element from the ring, `fromInteger()` to embed the natural numbers into the ring, `random()` to create a random element or `isCommutative()` to query if the ring is commutative.

The generic polynomial class `GenPolynomial` implements the `RingElem` interface and specifies that generic coefficients must be of type `RingElem`. In addition to the methods mandated by the interface, the `GenPolynomial` implements the methods like `leadingMonomial()` or `extend()` and `contract()` to transform the polynomial to 'bigger' or 'smaller' polynomial rings.

Polynomials are to be created via, respectively with, a polynomial factory `GenPolynomialRing`. In addition to the ring factory methods it defines for example a method to create random polynomials with parameters for coefficient size, number of terms, maximal degree and exponent vector density. The constructor for `GenPolynomialRing` takes parameters for a factory for the coefficients, the number of variables, the names for the variables and a term order object `TermOrder`.

For further details on the JAS types, interfaces and classes see [37, 29, 31, 33].

To get an idea of the interplay of the types, classes and object construction consider the following type

```
List<GenPolynomial<Product<Residue<BigRational>>>>
```

of a list of polynomials over a direct product of residue class rings modulo some polynomial ideal over the rational numbers. It arises in the computation of Gröbner bases over commutative regular rings, see [47, 53, 54].

$$R = \mathbb{Q}[x_1, \ldots, x_n], \quad S' = \Big(\prod_{\mathfrak{p} \in \mathrm{spec}(R)} R/\mathfrak{p} \Big)[y_1, \ldots, y_r].$$

To keep the example simple we will show how to generate a list L of polynomials in the ring

$$L \subset S = (\mathbb{Q}[x_0, x_1, x_2]/\mathrm{ideal}(F))^4[a, b].$$

The ring S is represented by the object in variable `fac` in the listing in Fig. 2. Random polynomials of this ring may look like the one shown in Fig. 3. The coefficients from $(\mathbb{Q}[x_0, x_1, x_2]/\mathrm{ideal}(F))^4$ are shown enclosed in braces `{}` in the form `i=polynomial`. I.e. the index `i` denotes the product component $i = 0, 1, 2, 3$ which reveals that the `Product` class is implemented using a sparse data structure. The list of F is printed after the 'rr =' together with the indication of the type of the residue class ring `ResidueRing` as polynomial ring in the variables `x0`, `x1`, `x2` over the rational numbers `BigRational` with graded lexicographical term order `IGRLEX`. The variables `a`, `b` are from the 'main' polynomial ring and the rest of Fig. 3 should be obvious.

The output in Fig. 3 is computed by the program from Fig. 2. Line number 1 defines the variable L of our intended type and creates it as an Java `Array-List`. Lines 2 and 3 show the creation of the base polynomial ring $\mathbb{Q}[x_0, x_1, x_2]$

```
1   List<GenPolynomial<Product<Residue<BigRational>>>> L
      = new ArrayList<GenPolynomial<Product<Residue<BigRational>>>>();
2   BigRational bf = new BigRational(1);
3   GenPolynomialRing<BigRational> pfac
      = new GenPolynomialRing<BigRational>(bf,3);
4   List<GenPolynomial<BigRational>> F
      = new ArrayList<GenPolynomial<BigRational>>();
5   GenPolynomial<BigRational> pp = null;
6   for ( int i = 0; i < 2; i++) {
7       pp = pfac.random(5,4,3,0.4f);
8       F.add(pp);
9   }
10  Ideal<BigRational> id = new Ideal<BigRational>(pfac,F);
11  id.doGB();
12  ResidueRing<BigRational> rr = new ResidueRing<BigRational>(id);
13  System.out.println("rr = " + rr);
14  ProductRing<Residue<BigRational>> pr
      = new ProductRing<Residue<BigRational>>(rr,4);
15  String[] vars = new String[] { "a", "b" };
16  GenPolynomialRing<Product<Residue<BigRational>>> fac
      = new GenPolynomialRing<Product<Residue<BigRational>>>(pr,2,vars);
17  GenPolynomial<Product<Residue<BigRational>>> p;
18  for ( int i = 0; i < 3; i++) {
19      p = fac.random(2,4,4,0.4f);
20      L.add(p);
21  }
22  System.out.println("L = " + L);
23  GroebnerBase<Product<Residue<BigRational>>> bb
      = new RGroebnerBasePseudoSeq<Product<Residue<BigRational>>>(pr);
24  List<GenPolynomial<Product<Residue<BigRational>>>> G = bb.GB(L);
25  System.out.println("G = " + G);
```

Fig. 2. Constructing algebraic objects

in variable `pfac`. In lines 4 to 9 a list F of random polynomials is constructed which will generate the ideal of the residue class ring. Lines 10 to 13 create a Gröbner basis for the ideal, setup the residue class ring `rr` and print it out. Line 14 constructs the regular ring `pr` as direct product of 4 copies of the residue class ring `rr`. The the final polynomial ring `fac` in the variables a and b is defined in lines 15 and 16. Lines 17 to 22 then generate the desired random polynomials, put them to the list L and print it out. The last lines 23 to 25 show the instantiation of a Gröbner base algorithm for regular coefficient rings `bb` and the computation of a Gröbner base G. `GroebnerBasePseudo` means the fraction free algorithm for coefficient arithmetic. To keep polynomials at a reasonable size, the primitive part of the polynomials is used. This requires gcd computations on `Product` objects, which is possible by our design, see 4.1.

```
rr = ResidueRing[ BigRational( x0, x1, x2 ) IGRLEX
      ( ( x0^2 + 295/336  ), ( x2 - 350/1593 x1 - 1100/2301 ) ) ]
L = [
   {0=x1 - 280/93 , 2=x0 * x1 - 33/23 } a^2 * b^3
 + {0=122500/2537649 x1^3 + 770000/3665493 x1^2
      + 14460385/47651409 x1 + 14630/89739 ,
      3=350/1593 x1 + 23/6 x0 + 1100/2301 } , ... ]
```

Fig. 3. Random polynomials from ring S

With this example we see that the software representations of rings snap together like 'LEGO blocks' to build up arbitrary structured rings. This concludes the introduction to JAS, further details can be found, as already mentioned, in [37, 29, 31, 33].

3   GCD Class Layout

In this section we discuss the overall design considerations for the implementation of a library for multivariate polynomial greatest common divisor (gcd) computations. We assume that the reader is familiar with the importance and the mathematics of the topic, presented for example in [28, 17, 10].

3.1   Design Overview

For the implementation of the multivariate gcd algorithm we have several choices

1. where to place the algorithms in the library,
2. which interfaces to implement, and
3. which recursive polynomial methods to use.

For the *first* item we could place the gcd algorithms into the class GenPolynomial, to setup an new class, or setup a new package. Axiom [24] places the the gcd algorithms directly into the abstract polynomial class (called *category*). For the type system this would be best, since having an gcd algorithm is a property of the multivariate polynomials. Most library oriented systems, place the gcd algorithms in a separate package in a separate class. This is to keep the code at a manageable size. In our implementation the class GenPolynomial consists of about 1200 lines of code, whereas all gcd classes (with several gcd implementations) consist of about 3200 lines of code. For other systems this ratio is similar, for example in MAS [38] or Aldes/SAC-2 [11]. The better maintainability of the code has led us to choose the separate package approach. The new package is called edu.jas.ufd. We leave a simple gcd() method, only for univariate polynomials over fields, in the class GenPolynomial.

The other two items are discussed in the following subsections.

3.2   Interface `GcdRingElem`

The *second* item, the interface question, is not that easy to decide. In our type
hierarchy (see Fig. 1) we would like to let `GenPolynomial` (or some sub-class)
implement `GcdRingElem` (an extension of `RingElem` by the methods `gcd()` and
`egcd()`) to document the fact that we can compute gcds. This is moreover required, if we want to use polynomials as coefficients of polynomials and want to
compute gcds in these structures. To take the content of such a polynomial we
must have the method `gcd()` available for the coefficients.

   As a resort, one could also extend `GenPolynomial` to let the sub-class implement `GcdRingElem`, for example

```
class GcdGenPolynomial<C extends GcdRingElem<C>>
     extends GenPolynomial<C>
     implements GcdRingElem<GcdGenPolynomial<C>>.
```

As we have noted in [30, 31], this is not possible, since sub-classes cannot implement the same interface as the superclass with different generic type parameters. I.e. `RingElem` would be implemented twice with different type parameters:
`GcdGenPolynomial` and `GenPolynomial`, which is not type-safe and so it is not
allowed in Java.

   Another possibility is to let `GenPolynomial` directly implement `GcdRingElem`.
Then we could, however, not guarantee that the method `gcd()` can always be
implemented. There can be cases, where `gcd()` will fail to exist and an exception
must be thrown. But we have accepted such behavior already with the method
`inverse()` in `RingElem`. If we chose this alternative and let `GenPolynomial`
implement `GcdRingElem`, eventually with `GcdRingElem` as coefficient types, we
would have to change nearly all existing classes. I.e. more than 100 coefficient
type restrictions must be be adjusted from `RingElem` to `GcdRingElem`.

   Despite of this situation we finally decided to let `RingElem` directly define
`gcd()`, but with no guarantees that it will not throw an exception. This requires
only 10 existing classes to additionally implement a `gcd()` method. Some throw
an exception, and for others (like fields) the gcd is trivially 1. So now `GcdRing-
Elem` is only a marker interface and `RingElem` itself defines `gcd()`.

3.3   Recursive Methods

The *third* item, the recursive polynomial methods question, is discussed in this
section.

   We have exercised some care in the definition of our interfaces to ensure, that
we can define recursive polynomials (see Fig. 1).

   First, the interface `RingElem` is defined as

```
RingElem<C extends RingElem<C>>.
```

So the type parameter C can (and must) itself be of type `RingElem`. With this
we can define polynomials with polynomials as coefficients

```
GenPolynomial<GenPolynomial<BigRational>>.
```

In the applications implemented so far we did not make much use of this feature. However, there are many algebraic algorithms which are only meaningful in a recursive setting, for example greatest common divisors, resultants or factorization of multivariate polynomials.

If we use our current implementation of `GenPolynomial`, we observe, that our type system will unfortunately lead to code duplication. Consider the greatest common divisor method `gcd()` with the specification

```
GenPolynomial<C> gcd( GenPolynomial<C> P, GenPolynomial<C> S )
```

This method will be a driver for the recursion. It will check if the number of variables in the polynomials is one, or if it is greater than one. In the first case, a method for the recursion base case of univariate polynomials, must be called

```
GenPolynomial<C> baseGcd( GenPolynomial<C> P, S ).
```

In the second case, the polynomials have to be converted to recursive representation and a method for the recursion case must be called

```
GenPolynomial<GenPolynomial<C>>
recursiveUnivariateGcd( GenPolynomial<GenPolynomial<C>> P, S ).
```

The type of the parameters for `recursiveUnivariateGcd()` is univariate polynomials with (multivariate) polynomials as coefficients. The Java code for `base-Gcd()` and `recursiveUnivariateGcd()` is mostly the same, but because of the type system, the methods must have different parameter types. Further, by type erasure for generic parameters during compilation, they must also have different names.

3.4   Conversion of Representation

In the setting described in the previous section, we need methods to convert between the distributed and recursive representation. In the class `PolyUtil` we have implemented two static methods for this purpose. In the following, assume that C extends `RingElem<C>`.

The first method converts a distributed polynomial A to a recursive polynomial in the polynomial ring defined by the ring factory `rf`.

```
GenPolynomial<GenPolynomial<C>>
  recursive( GenPolynomialRing<GenPolynomial<C>> rf,
             GenPolynomial<C> A )
```

In the method `gcd()` the recursive polynomial ring `rf` will be an univariate polynomial ring with multivariate polynomials as coefficients.

The second method converts a recursive polynomial B to a distributed polynomial in the polynomial ring defined by the ring factory `dfac`.

```
GenPolynomial<C>
  distribute( GenPolynomialRing<C> dfac,
              GenPolynomial<GenPolynomial<C>> B)
```

We have not yet studied the performance implications of many back and forth conversions between these two representations. For a sketch of a recursive polynomial representation in Java see the appendix in [32].

4   GCD Implementations

In this section we present the most important algorithmic versions for gcd computation. An overview of the classes is given in Fig. 4. The class relations are modeled after the *template method* design pattern, see [16].

We start with an interface `GreatestCommonDivisor`. It defines the method names for a ring with gcd algorithm. First there is the method `gcd()` itself, together with the method `lcm()` to compute the least common multiple of two polynomials. With the help of `gcd()` the algorithms for the content `content()` and the primitive part `primitivePart()` computation can be implemented. With the help of a polynomial derivative, the square-free part `squarefree-Part()` and the square-free factors `squarefreeFactors()` can be implemented. Finally the resultant of two polynomials `resultant()` is defined. It doesn't use

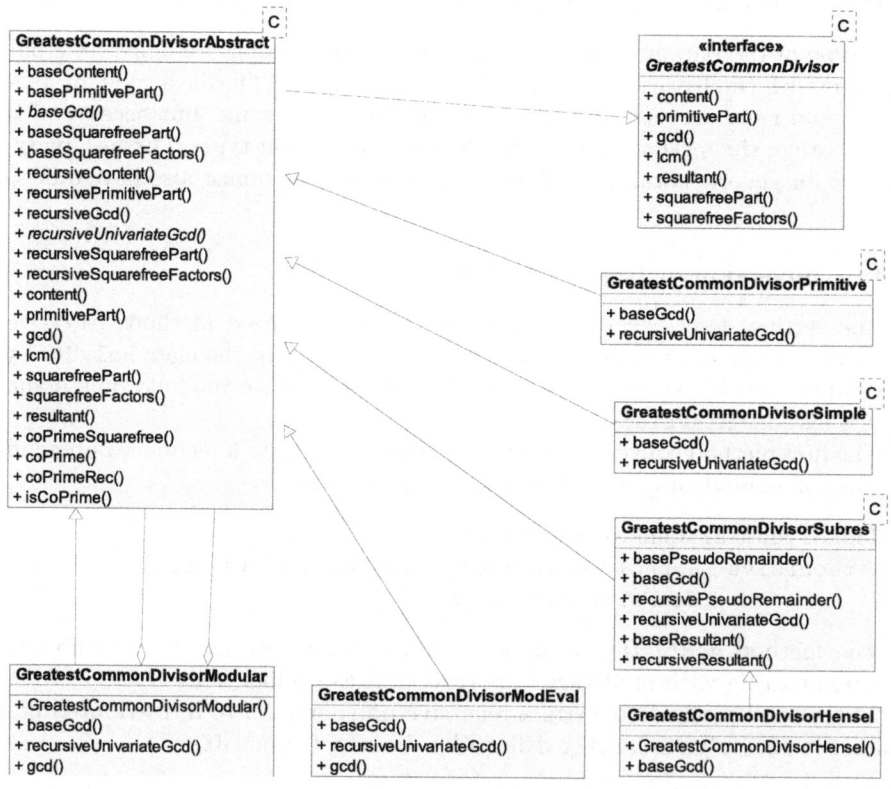

Fig. 4. Greatest common divisor classes

the `gcd()` implementation, but is implemented like gcd. Since ADG, the interface has been enhanced by methods `coPrime()` to compute lists of co-prime polynomials from given lists of polynomials.

The abstract super class for the implementations is called `GreatestCommon-DivisorAbstract`. It implements nearly all methods defined in the `GreatestCommonDivisor` interface. The abstract methods are `baseGcd()` and `recursiveUnivariateGcd()`. The method `gcd()` first checks for the recursion base, and eventually calls `baseGcd()`. Otherwise it converts the input polynomials to recursive representation, as univariate polynomials with multivariate polynomial coefficients, and calls method `recursiveUnivariateGcd()`. The result is then converted back to the distributed `GenPolynomial` representation. The method `recursiveGcd()` can be used for arbitrary recursive polynomials (not only univariate ones). The input polynomials are converted to distributed form, then `gcd()` is called and finally the result polynomial is converted back to the given recursive form.

The concrete implementations come in two flavors. The first flavor implements only the methods `baseGcd()` and `recursiveUnivariateGcd()`, using the setup provided by the abstract super class. The second flavor directly implements `gcd()` without providing `baseGcd()` and `recursiveUnivariateGcd()`. However, these versions are only valid for the JAS coefficient classes `BigInteger` or `ModInteger`.

4.1 Polynomial Remainder Sequences

Euclid's algorithm to compute gcds by taking remainders is good for the integers and works for polynomials. For polynomials it is, however, inefficient due to the *intermediate expression explosion*, i.e. the base coefficients and the degrees of the coefficient polynomials can become extremely large during the computation. The result of the computation can nevertheless be surprisingly small, for example 1.

To overcome this coefficient explosion, the intermediate remainders are simplified according to different strategies. The sequence of intermediate remainders is called *polynomial remainder sequence* (PRS), see [17, 28, 41].

The maximum simplification is achieved by the *primitive PRS*. In this case, from the remainder in each step, the primitive part is taken. However, the computation of primitive parts can be very expensive too, since a gcd of all the coefficients of the remainder must be computed. The primitive PRS gcd is implemented by the class `GreatestCommonDivisorPrimitive`.

The best known remainder sequence is the *sub-resultant PRS*. In this algorithm, the coefficients of the remainder are divided by a polynomial, which can be computed at nearly no additional cost. The simplification gained, is not so good, as with the primitive PRS, but the coefficients are kept at nearly the same size, and it doesn't cost the recursive gcd computations. The sub-resultant PRS gcd is implemented by the class `GreatestCommonDivisorSubres`.

The original euclidean algorithm is implemented by the class `GreatestCommon-DivisorSimple`. Here, no simplifications of the remainder are applied, except if the base coefficient ring is a field. In this case the remainder polynomials are made monic, i.e. the leading base coefficient is multiplied by its inverse to make it equal to 1. This PRS is also called the *monic PRS*.

The PRS algorithm implementations make no use of specific properties of the coefficients, only a `gcd()` method is required to exist. So they can be used in very general settings, as shown in Sect. 2, Fig. 2. In this example the class `Residue` implements `gcd()` as gcd of the canonical residue class representatives and the class `Product` implements `gcd()` as component wise gcds.

4.2   Modular Methods

For the special case of integer coefficients, there are even more advanced techniques. First the base integer coefficients are mapped to elements of finite fields, i.e. to integers modulo a prime number. If this mapping is done with some care, then the original polynomials can be reconstructed by the *Chinese remainder theorem*, see [17]. Recall, that the algorithm proceeds roughly as follows:

1. Map the coefficients of the polynomials modulo some prime number p. If the mapping is not 'good', choose a new prime and continue with step 1.
2. Compute the gcd over the modulo p coefficient ring. If the gcd is 1, also the 'real' gcd is one, so return 1.
3. From gcds modulo different primes reconstruct an approximation of the gcd using Chinese remaindering. If the approximation is 'correct', then return it, otherwise, choose a new prime and continue with step 1.

The algorithm is implemented by the class `GreatestCommonDivisorModular`. Note, that it is no longer generic, since it explicitly requires `BigInteger` coefficients. It extends the super class `GreatestCommonDivisorAbstract<BigInteger>` with fixed type parameter. The gcd in step 2 can be computed by a monic PRS, since the coefficients are from a field, or by the algorithm described next.

The Chinese remaindering can not only be applied with prime numbers, but also with prime polynomials. To make the computation especially simple, one chooses linear univariate prime polynomials. This amounts then to the evaluation of the polynomial T at the constant term of the prime polynomial, i.e. $T \bmod (x - a) = T(a)$. By this evaluation also a variable of the input polynomial disappears. With this method the recursion is done by removing one variable at a time, until the polynomial is univariate. Then the monic PRS is used to compute the gcd. In the end, step by step the polynomials are reconstructed by Chinese remaindering until all variables are present again. This algorithm is implemented by the class `GreatestCommonDivisorModEval`. As above, it is no longer generic, since it requires `ModInteger` coefficients. It extends the super class `Greatest-CommonDivisorAbstract<ModInteger>` with fixed type parameter.

There is another modular algorithm using the *Hensel lifting*. In this algorithm, the reconstruction of the real gcd from a modular gcd is done modulo powers p^e of a prime number p. This algorithm is implemented for univariate polynomials in class `GreatestCommonDivisorHensel`. It extends `GreatestCommonDivisor-Subres<BigInteger>` so that for multivariate polynomials first the sub-resultant algorithm is used and then in the univariate case Hensel lifting is performed. In case, the greatest common divisor is not co-prime to the cofactor the algorithm falls back to the univariate sub-resultant algorithm.

4.3   GCD Factory

All the presented algorithms have pros and cons with respect to the computing time. So which algorithm to choose? This question can seldom by answered by mere users of the library. So it is the task of the implementers to find a way, to choose the optimal implementation. We do this with the class `GCDFactory`, which employs the factory pattern, see [16], to construct a suitable gcd implementation. Its definition is

```
GreatestCommonDivisor<C> GCDFactory.<C>getImplementation(fac).
```

The static method `getImplementation()` constructs a suitable gcd implementation, based on the given type parameter C. This method has a coefficient factory parameter `fac`. For, say, `BigInteger` coefficients we would compute the gcd of two polynomials of this type by

```
GreatestCommonDivisor<BigInteger> engine
  = GCDFactory.getImplementation(fac);
c = engine.gcd(a,b);
```

Here, `engine` denotes a variable with the gcd interface type, which holds a reference to an object implementing the gcd algorithm. An alternative to `get-Implementation()` is a method `getProxy()`, with the same signature,

```
GreatestCommonDivisor<C> GCDFactory.<C>getProxy(fac),
```

which returns a gcd proxy class. It is described in the next section.

Currently four versions of `getImplementation()` are implemented for the coefficient types `ModInteger`, `BigInteger`, `BigRational` and the most general `GcdRingElem<C>`. The selection of the respective version takes place at compile time. I.e. depending on the specified type parameter C, different methods are selected. The coefficient factory parameter `fac` is used at run-time to check if the coefficients are from a field, to further refine the selection.

```
static GreatestCommonDivisor<ModInteger>
       getImplementation( ModInteger fac )

static ... GreatestCommonDivisor<BigInteger> ...

static <C extends GcdRingElem<C>> GreatestCommonDivisor<C>
       getImplementation( RingFactory<C> fac ).
```

To use non-static methods, it would be necessary, to instantiate `GCDFactory` objects for each type separately. The gcd factories are, however, difficult to implement when the concrete type parameter, for example `BigInteger`, is not available. To overcome this problem, the last method must be able to handle also the first three cases at run-time.

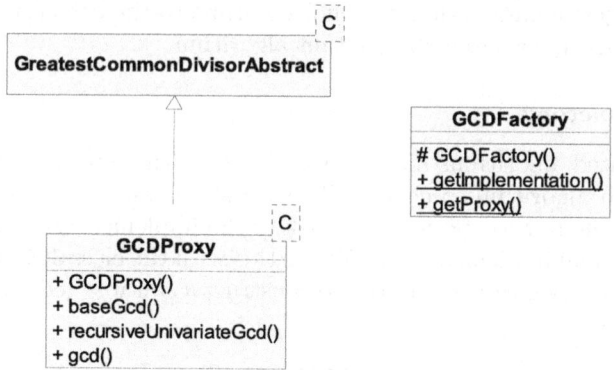

Fig. 5. GCD proxy class

4.4   GCD Proxy

The selection of a gcd algorithm with the class `GCDFactory` is not optimal for all kinds of input polynomial shapes. The modular algorithms are generally the best, but for some particular polynomials the PRS algorithms perform better. There are many investigations on the complexity of the gcd algorithms, see [17] and the references therein. The complexity bounds can be calculated depending on further polynomial shape properties:

- the *size* of the coefficients,
- the *degrees* of the polynomials, and
- the *number* of variables,
- the *density* or *sparsity* of polynomials, i.e. the relation between the number of non-zero coefficients to the number of all possible coefficients for the given degrees,
- and the *density* of the exponents.

However, the determination of all these properties, together with the estimation of the execution time, require a substantial amount of computing time, which can not be neglected. Moreover, the estimations are often worst case upper bound estimates which are of little value in practice. These limitations have been addressed by the concept of speculative parallelism in [48].

In the time of multi-core CPU computers, we can do better than the precise complexity case distinctions: we can compute the gcd with two (or more) different implementations in parallel. Then we return the result from the fastest algorithm, and cancel the other still running one. Java provides all this in the package `java.util.concurrent`. A sketch of the gcd proxy class `GCDProxy` and the method gcd follows.

```
final GreatestCommonDivisor<C> e1, e2;
protected ExecutorService pool; // set in constructor
GenPolynomial<C> gcd(final GenPolynomial<C> P, ... S) {
  List<Callable<GenPolynomial<C>>> cs = ...;
```

```
cs.add( new Callable<GenPolynomial<C>>() {
        public GenPolynomial<C> call() {
            return e1.gcd(P,S);
        }
    }
);
cs.add( ... e2.gcd(P,S); ... );
return pool.invokeAny( cs );
}
```

The variables e1 and e2 hold gcd implementations, and are set in the constructor of GCDProxy, for example in the method getProxy() of GCDFactory. pool is an ExecutorService, which provides the method invokeAny(). It executes all Callable objects in the list cs in parallel. After the termination of the fastest call() method, the other, still running methods are canceled. The result of the fastest call() is then returned by the proxy method gcd(). The polynomials P and S are the actual parameters of the proxy method gcd().

With this proxy gcd implementation, the computation of the gcds is as fast as the computation of the fastest algorithm for a given coefficient type, and the run-time shape of the polynomials. This is true, if there are more than one CPU in the computer and the other CPU is idle. If there is only one CPU, or the second CPU is occupied by some other computation, then the computing time of the proxy gcd method is not worse than two times the computing time of the fastest algorithm. This will in most cases be better, since the computing times of the different algorithms with different polynomial shapes differ in general by more than a factor of two.

For the safe forced termination of the second running computation a new Java exception PreemptingException had to be introduced. It is checked in the construction of new polynomials via the class PreemptStatus which is queried during polynomial ring construction.

5   GCD Performance

In this section we report on some performance measurements for the different algorithms. Our tests are not intended to be comprehensive like in [39] or the references contained therein. We just want to make sure that our implementation qualitatively matches the known weaknesses and strengths of the algorithms.

5.1   Relative Algorithm Performance

In this sub-section we compare our implemented algorithms against each other. We generate three *random* polynomials a, b, c, with the shape parameters given in the respective figure. Then we compute the gcd of ac and bc, let $d = \gcd(ac, bc)$ and check if $c|d$. The random polynomial parameters are: the number of variables r, the size of the coefficients 2^k, the number of terms l, the maximal degree e, and the density of exponent vectors q. The maximal degrees of the polynomials

degrees, e	s	p	sr	ms	me
a=7, b=6, c=2	23	23	36	1306	2176
a=5, b=5, c=2	12	19	13	36	457
a=3, b=6, c=2	1456	117	1299	1380	691
a=5, b=5, c=0	508	6	6	799	2

s = simple, p = primitive, sr = sub-resultant, ms = modular simple monic, me = modular evaluation. `random()` parameters: r = 4, k = 7, l = 6, q = 0.3,

Fig. 6. PRS and modular gcd performance

degrees, e	sr	ms	me
a=5, b=5, c=0	3	29	27
a=6, b=7, c=2	181	695	2845
a=5, b=5, c=0	235	86	4
a=7, b=5, c=2	1763	874	628
a=4, b=5, c=0	26	1322	12

sr = sub-resultant, ms = modular simple monic, me = modular evaluation. `random()` parameters: r = 4, k = 7, l = 6, q = 0.3,

Fig. 7. Sub-resultant and modular gcd performance

degrees, e	time	algorithm
a=6, b=6, c=2	3566	subres
a=5, b=6, c=2	1794	modular
a=7, b=7, c=2	1205	subres
a=5, b=5, c=0	8	modular

BigInteger coefficients, winning algorithm: subres = sub-resultant, modular = modular simple monic. `random()` parameters: r = 4, k = 24, l = 6, q = 0.3,

Fig. 8. Proxy gcd performance, integer

degrees, e	time	algorithm
a=6, b=6, c=2	3897	modeval
a=7, b=6, c=2	1739	modeval
a=5, b=4, c=0	905	subres
a=5, b=5, c=0	10	modeval

ModInteger coefficients, winning algorithm: subres = sub-resultant, modeval = modular evaluation. `random()` parameters: r = 4, k = 6, l = 6, q = 0.3,

Fig. 9. Proxy gcd performance, modular integer

are shown in the left column. The computing times are in milliseconds on one AMD 1.6 GHz CPU, with JDK 1.5 and server VM.

Comparisons in Figs. 6 and 7 show differences of factors of 10 to 100 for the different algorithms and different polynomial shapes. The comparisons for the

random parameters	jas(1)	jas(2)	map	map n
r=2, k=1, l=3, d=4, q=0.6	15	2	28	20
r=2, k=2, l=4, d=8, q=0.5	6	6	28	20
r=2, k=3, l=5, d=12, q=0.4	4	3	28	20
r=2, k=4, l=6, d=16, q=0.3	6	8	28	40
r=2, k=5, l=7, d=20, q=0.2	5	5	28	44
r=2, k=6, l=8, d=24, q=0.1	3	1	32	44

Polynomials in 2 variables, $c = 1$.

Fig. 10. Performance comparison to other CAS (1)

random parameters	jas(1)	jas(2)	map	map n
r=6, k=1, l=3, d=4, q=0.6	17	6	8	24
r=6, k=2, l=4, d=8, q=0.5	11	15	12	36
r=6, k=3, l=5, d=12, q=0.4	19	19	12	44
r=6, k=4, l=6, d=16, q=0.3	38	29	16	48
r=6, k=5, l=7, d=20, q=0.2	20	19	16	48
r=6, k=6, l=8, d=24, q=0.1	10	6	16	48

Polynomials in 6 variables, $c = 1$.

Fig. 11. Performance comparison to other CAS (2)

random parameters	jas(1)	jas(2)	map	map n
r=9, k=5, l=3, d=4, q=0.6	22	15	24	20
r=9, k=10, l=4, d=8, q=0.5	29	34	28	44
r=9, k=15, l=5, d=12, q=0.4	249	36	32	52
r=9, k=20, l=6, d=16, q=0.3	25	61	32	56
r=9, k=25, l=7, d=20, q=0.2	21	19	36	60
r=9, k=30, l=8, d=24, q=0.1	10	9	40	60

Polynomials in 9 variables, with constant greatest common divisor $(c = 1)$.

Fig. 12. Performance comparison to other CAS (3)

GCDProxy are contained in Figs. 8 and 9. We see that for each run an other algorithm may be the fastest, which exemplifies the usefulness of the proxy gcd approach.

5.2   Comparison with other CAS

We generate three *random* polynomials a, b, with the shape parameters given in the respective figure. Polynomial c is either constant $= 1$ or a random polynomial with 2 to 3 terms. We compute the gcd of ac and bc, let $d = \gcd(ac, bc)$ and check if $c|d$. The random polynomial parameters shown in the left column are: the number of variables r, the size of the coefficients 2^k, the number of terms l, the maximal degree e, and the density of exponent vectors q.

All computing times are in milliseconds on a 32-bit computer with 16 Intel XEON hyper-threading CPUs running at 2.7 GHz. Measurements are with JDK 1.5 and the 32-bit server JVM. The JDK 1.6 is known to be faster about $10 - 20\%$, but has not been used for this tests. We compare JAS to Maple version 9.5 and version 11.0. In Figs. 10, 11, 12, 13 and 14 these versions are named "map" and "map n" respectively. The columns named "jas(1)" and "jas(2)" are timings for the first, respectively the second run in the same JVM instance. From comparisons between Maple and other computer algebra systems (as for example from [39]) one may then draw some conclusions about the comparison between JAS and the other systems. But such comparisons always depend on the system version and the available computer infrastructure and should be interpreted with caution. The Maple timing code is

```
st:=time():
d:=gcd(ac,bc):
time()-st:
print("maple time = ", st, " gcd = ", d):
```

and the JAS (jython) timing code is

```
tt = System.currentTimeMillis();
g = R.gcd(ac,bc);
tt = System.currentTimeMillis() - tt;
tt2 = System.currentTimeMillis();
g2 = R.gcd(ac,bc);
tt2 = System.currentTimeMillis() - tt2;
print "jas gcd time =", tt, " (", tt2, ") milliseconds"
```

`R.gcd(ac,bc)` is the gcd method of the respective polynomial ring. It uses the GCDProxy for the `BigInteger` coefficient ring, that is, the sub-resultant algorithm and the modular algorithm run in parallel. `tt` is the time for the first run and `tt2` is the time for the second run.

In Figs. 10, 11 and 12 with a constant greatest common divisor we see similar computing times for JAS and Maple. In Figs. 13 and 14 with a non-constant greatest common divisor we see varying results. For certain random parameters

random parameters	jas(1)	jas(2)	map	map n
r=2, k=35, l=9, d=8, q=0.6	79	32	20	28
r=2, k=70, l=12, d=16, q=0.5	37	32	24	56
r=2, k=105, l=15, d=24, q=0.4	676	328	24	64
r=2, k=140, l=18, d=32, q=0.3	18	15	28	72
r=2, k=175, l=21, d=40, q=0.2	22	21	32	72
r=2, k=210, l=24, d=48, q=0.1	15	14	36	76

Polynomials in 2 variables, $c \neq 1$, with 2 - 3 terms.

Fig. 13. Performance comparison to other CAS (4)

JAS is slower than Maple (see Fig. 13) but in Fig. 14 we see that Maple 9.5 is slower than JAS, but JAS is comparable to Maple 11.0. Note, that JAS runs two algorithms in parallel, where for Maple it is not known to us, if multiple CPUs are used if available.

5.3 Application Performance

With the gcd implementation we re-factored the rational function class `Quotient` in package `edu.jas.application`. To reduce quotients of polynomials to lowest terms the gcd of the nominator and denominator is divided out. Due to the 'LEGO block' design, the new ring can be used also as coefficients for polynomials. So the Gröbner bases implementation can directly be used for these rings (without recompilation). We compare the computing time with the system MAS [38], since we are familiar with the implementation details in it. Other systems where a computation in such a setting is possible will be compared later.

We studied the performance on the examples from Raksanyi [6] and Hawes2 [5, 19]. In the first table in Fig. 15 we compared the computation with MAS [38]. The computing times are in milliseconds on an AMD 1.6 GHz CPU, with JDK 1.6 for JAS. For JAS we compute the same Gröbner base two times, the time for the second run is enclosed in parenthesis. We see, that for the first run there is considerable time spend in JVM 'warm-up', i.e. code profiling and just-in-time compilation. For the client-VM, second run, the timing for the Raksanyi example is the same magnitude as for MAS. In case of the server-VM we see, that even in the second run, the JVM tries hard to find more optimizations. In this case the computing times are less than half of the first run, but slower than the MAS timings. For the graded Hawes2 example the computing times are in an equal range. The same example with lexicographic term order shows more differences: the computation with JAS is about 2 times faster than MAS, even in the first run and further improves in the second run.

Note, that in all examples the server JVM is slightly slower than the client JVM since the additional optimization of the server JVM can not be amortized in this short running examples. In a second or subsequent runs the timings generally improve due to optimizations by just-in-time compilation.

The second table in Fig. 15 shows the gcd algorithm count for the Hawes2 example. Here most of the time the sub-resultant algorithm was fastest. But

random parameters	jas(1)	jas(2)	map	map n
r=9, k=5, l=3, d=4, q=0.6	61	55	16	16
r=9, k=10, l=4, d=8, q=0.5	792	293	20	44
r=9, k=15, l=5, d=12, q=0.4	147	94	24	52
r=9, k=20, l=6, d=16, q=0.3	15	17	2096	64
r=9, k=25, l=7, d=20, q=0.2	17	16	2100	68
r=9, k=30, l=8, d=24, q=0.1	16	14	2100	68

Polynomials in 9 variables with $c \neq 1$, with 2 - 3 terms.

Fig. 14. Performance comparison to other CAS (5)

since this was run on a one CPU computer it only shows the preference of the JVM scheduler for the first started algorithm (compare Sect. 4.4).

The Hawes2 example [5, 19] is used in Fig. 16 to compare different gcd algorithms for the rational function coefficients. The computing times are in milliseconds on a 16 CPU 64-bit AMD Opteron 2.6 GHz, with JDK 1.5 and server JVM. The first three lines show computations with the GCDProxy running two algorithms in parallel and the remaining lines show the times for only one gcd algorithm. The counts vary since two implementations may terminate at nearly the same time, so eventually the discarded result is also counted as success. In this example the integer coefficients in the rational function coefficients are smaller than 2^{28}. So the sub-resultant algorithm alone is faster than the modular algorithms and the modular computations with primes slightly less than 2^{59} are slower than the computations modulo small primes. The modular evaluation algorithm is faster than the monic PRS algorithm in the recursion.

The comparison is not statistically rigorous in the sense of [18] as we have done only two runs of the methods.

6   GCD Evaluation

In the following we summarize the pros and cons of the new algorithms. We start with the problems and end with the positive aspects.

- We are using a distributed representation with conversions to recursive representation on demand. It is not investigated how costly these many conversions between distributed and recursive representation are. There are also manipulations of ring factories to setup the correct polynomial ring for recursions. Compared to MAS with a recursive polynomial representation, the timings in Fig. 15 indicate that the conversions are indeed acceptable.
- The class ModInteger for polynomial coefficients is implemented using BigInteger. Systems like Singular [21], MAS [38] and Aldes/SAC-2 [11], use ints for modular integer coefficients. This can have great influence on the

example	MAS	JAS, clientVM	JAS, serverVM
Raksanyi, G	50	311 (53)	479 (205)
Raksanyi, L	40	267 (52)	419 (198)
Hawes2, G	610	528 (237)	1106 (1351)
Hawes2, L	26030	9766 (8324)	11061 (5966)

time in milliseconds for Gröbner base examples, Term order: G = graded, L = lexicographical, timings in parenthesis are for second run.

example/algorithm	Subres	Modular
Hawes2, G	1215	105
Hawes2, L	4030	125

Fig. 15. GB with rational function coefficients

gcd algorithm	time	count of first	count of second
Subres and Modular $p < 2^{28}$	5799	3807	2054
Subres and Modular $p < 2^{59}$	10817	3682	2100
Modular $p < 2^{28}$ and Subres	5662	2423	3239
Subres	5973		
Modular $p < 2^{28}$ with ModEval	21932		
Modular $p < 2^{28}$ with Monic	27671		
Modular $p < 2^{59}$ with Monic	34732		
Modular $p < 2^{59}$ with ModEval	24495		

time in milliseconds for Hawes2 lex Gröbner base example, first, second = count for respecive algorithm, p shows the size of the used prime numbers.

Fig. 16. Gcd on rational function coefficients

computing time. However, JAS is with this choice able to perform computations with arbitrary long modular integers. The right choice of prime size for modular integers is not yet determined. We experimented with primes of size less than `Long.maxValue` and less than `Integer.maxValue` which could eventually be exploited in `BigInteger`.

- The bounds used to stop iteration over primes, are not yet state of the art. We currently use the bounds found in [11]. The bounds derived in [10] and [17] are not yet incorporated. However, we try to detect factors by exact division, early.
- The univariate polynomials and methods are not separate implementations tuned for this case. I.e. we simply use the multivariate polynomials and methods with only one variable.
- There are no methods for extended gcds and half extended gcds implemented for multivariate polynomials yet. Better algorithms for the gcd computation of lists of polynomials are not yet implemented.
- For generic polynomials, such as `GenPolynomial<GenPolynomial<C>>` the gcd method can not be used. This sounds strange, but the coefficients are itself polynomials and so do not implement a multivariate polynomial gcd. In this case the method `recursiveGcd` must be called explicitly.
- We have designed a clean interface `GreatestCommonDivisor` with only the most useful methods. These are `gcd()`, `lcm()`, `primitivePart()`, `content()`, `squarefreePart()`, `squarefreeFactors()`, `resultant()` and `coPrime()`.
- The generic algorithms work for all implemented polynomial coefficients from (commutative) fields. The PRS implementations can be used in very general settings, as exemplified in Sects. 2 and 4.1.
- The abstract class `GreatestCommonDivisorAbstract` implements the full set of methods, as required by the interface. Only two methods must be implemented for the different gcd algorithms. The abstract class should eventually, be re-factored to provide an abstract class for PRS algorithms and an abstract class for the modular algorithms.

– The gcd factory allows non-experts of computer algebra to choose the right algorithm for their problem. The selection is first based on the coefficient type at compile time and more precisely at the field property at run-time. For the special cases of BigInteger and ModInteger coefficients, the modular algorithms can be employed. This factory approach contrasts the approach taken in [42] and [49] to provide programming language constructs to specify the requirements for the employment of certain implementations. These constructs would then direct the selection of the right algorithm, some at compile time and some at run-time.

– The new proxy class with gcd interface provides effective selection of the fastest algorithms at run-time. This is achieved at the cost of a parallel execution of two different gcd algorithms. This could waste maximally the time for the computation of the fastest algorithm. I.e. if two CPUs are working on the problem, the work of one of them is discarded. In case there is only one CPU, the computing time is maximally two times of the fastest algorithm.

7 Factorization

In this section we present a first working implementation of polynomial factorization. The mathematical background for polynomial factorization can be found for example in [28, 27, 17, 10], see also [20] for up-to-date references on the topic. Our goal is to have a working implementation for polynomial factorization over various coefficient rings to study suitable class layouts. In the future we will implement faster algorithms like the Berlekamp algorithm for the factorization over integers modulo a prime number or multivariate Hensel lifting for the multivariate polynomial factorization case.

7.1 Class Layout

Figure 17 shows the class layout of an interface, two abstract and several concrete classes. The interface Factorization defines the most useful factorization methods. The method factors() computes a complete factorization with no further preconditions and returns a SortedMap, which maps polynomials to the exponents of the polynomials occurring in the factorization. The method factorsSquarefree() factors a square-free polynomial. It returns a list of polynomials since the exponents will all be 1. Method factorsRadical() computes a complete factorization, but returns a list of polynomials, that is, all exponents are removed. Methods isIrreducible(), isReducible() and isSquarefree() test the respective properties for a polynomial. isFactorization() tests if a given map or list is actually a factorization for a given polynomial. For convenience there are methods squarefreePart() and squarefreeFactors() which delegate the computation to a suitable greatest common divisor engine.

The abstract class FactorAbstract implements all of the methods specified in the interface. Only baseFactorSquarefree() for the factorization of a square-free univariate polynomial is declared abstract and must be implemented for each coefficient ring. For multivariate polynomials Kronecker's algorithm is used to reduce this case to a univariate problem and to reassemble multivariate factors from univariate ones. This algorithm is not particularly fast and will be accompanied by a multivariate Hensel algorithm in the future.

7.2   Modular, Integer and Rational Coefficients

Class `FactorModular` implements `baseFactorSquarefree()` for polynomials with coefficients of integers modulo a prime number. To this end it uses the methods `distinctDegreeFactor()` and `equalDegreeFactor()` to first factor the polynomial into distinct degree parts and then to factor these parts into equal degree factors. Due to the required computation of high powers the method `modPower()` had to be implemented also for `BigInteger` powers, since normal `long` integers overflowed to easily. In the future these methods will be accompanied by a Berlekamp algorithm.

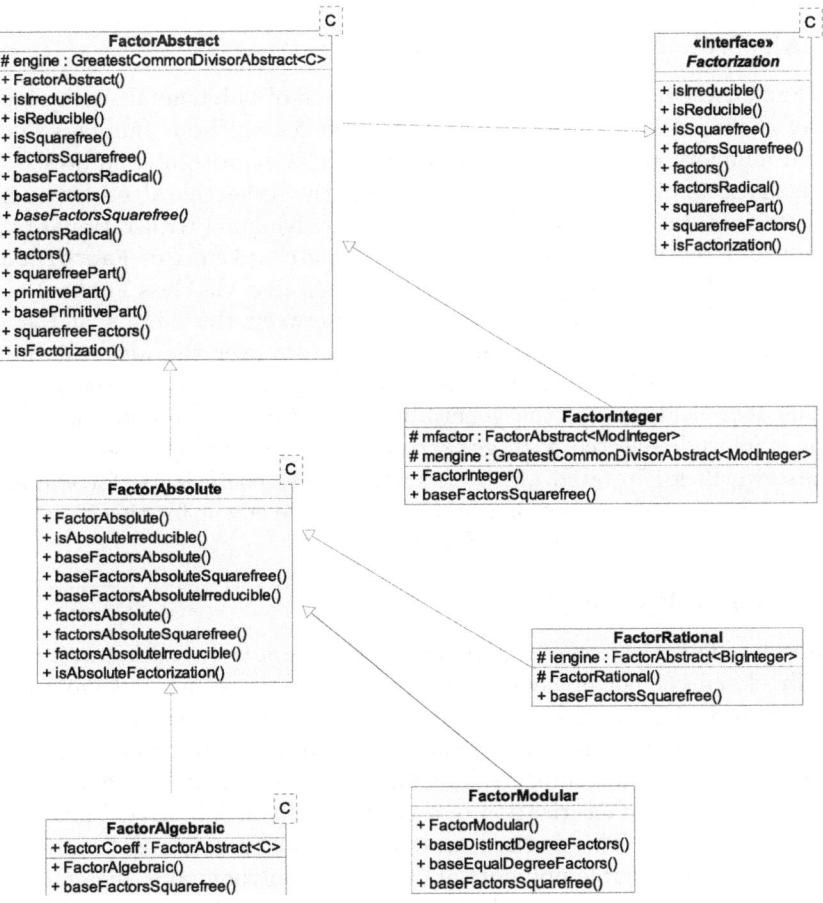

Fig. 17. Factorization classes and interfaces

Factorization of polynomials with integer coefficients is implemented in class `FactorInteger` and method `baseFactorSquarefree()`. In the algorithm of Zassenhaus, a polynomial is first factored modulo 'good' primes p, then the factors are lifted via Hensel to coefficients modulo p^e. The true factors over the integers are searched with combinatorial factor search. This implementation is experimental and has not yet reached a satisfying state. The factorization modulo a prime number is currently done with three small primes (starting with 3 until good primes are found) followed by one bigger prime number less than 2^{32}, starting with $2^{28} - 57$ and decreasing. From these four factorizations the one with the least number of factors is chosen for lifting. The modulo p factorizations will be done in parallel in the future.

Factorization of polynomials with rational number coefficients is implemented in class `FactorRational`. The algorithm simply clears denominators and uses factorization over integers. The leading coefficient of the input polynomial is attached to the first factor.

7.3 Algebraic Number Coefficients

Class `FactorAlgebraic` implements factorization of polynomials with algebraic number coefficients. The implementation works for algebraic numbers over the rational numbers and over modular integers. The algorithm computes a (suitable) norm of a polynomial to be factored over the respective algebraic numbers, then it factors this norm. Since the norm is a polynomial with rational number or modular integer coefficients the classes `FactorRational` or `FactorModular` will be used. The required implementation is selected via class `FactorFactory` (see below). The greatest common divisors between the factors of the norm and the given polynomial will then be its factors over the algebraic number field. Since the greatest common divisors have to be computed over an algebraic number field, one of the generic implementations presented in Sect. 4.1 will be employed (selected with `getProxy()` of `GCDFactory`). The implementation also works for iterated algebraic number field coefficients, for example in $\mathbb{Q}(i)(\sqrt{2})[x]$. Of corse, computing first a primitive element for this iterated field extension would speedup the computation.

7.4 Absolute Factorization

The absolute factorization of a polynomial, that is, a factorization over an algebraically closed coefficient field, is implemented in class `FactorAbsolute`. The implementation has been tested for an algebraically closed field over the rational numbers, but it should eventually work also for an algebraically closed field over prime modular integers. The main method is `baseFactorsAbsolute-Irreducible` for the factorization of a polynomial which is irreducible over the given coefficient field. This method constructs a field extension of the ground field from the given irreducible polynomial. The given polynomial is then factored in this algebraic field extension with the algorithms from class `FactorAlgebraic`.

This class contains further methods to compute polynomials which satisfy the preconditions of the main method. Method `isAbsoluteIrreducible` tests if a polynomial is absolutely irreducible by computing an absolute factorization, but only if the polynomial is irreducible. The methods `baseFactors-Absolute()` and `baseFactorsAbsoluteSquarefree()` are just used to compute irreducible factors which are then absolutely factored by `baseFactorsAbsolute-Irreducible()`. The methods for multivariate polynomials `factorsAbsolute()` and `factorsAbsoluteSquarefree()` are also only used to compute irreducible factors which are then absolutely factored by `factorsAbsoluteIrreducible()`. This method works by substituting random numbers for the variables to finally obtain a univariate squarefree polynomial. An irreducible factor of smallest degree of this polynomial is then used to construct a field extension of the ground field. The given multivariate polynomial is then factored over this algebraic number field and yields the desired absolute factorization.

Interestingly, class `FactorAbsolute` is itself abstract and extends abstract class `FactorAbstract`. To make it instantiable `FactorAbsolute` is extended by class `FactorRational`, class `FactorModular` and class `FactorAlgebraic`. In this way we enhance the classes for factoring over rational, modular prime numbers and algebraic extensions thereof with absolute factorization methods. An interface for the absolute factorization methods is not yet implemented. Note, there are algorithms to reduce the degree of the algebraic field extensions which are currently not implemented, but are planed for implementation in future work.

7.5 Factor Factory

For the selection of suitable factorization implementation we provide a factory named `FactorFactory`. Its method `getImplementation` selects an implementation based on the supplied polynomial coefficient factory. There is little choice at the moment, as there is only one algorithm implemented for each factoring problem. This will change as more and more alternative factoring methods become available.

As mentioned already, there will be a Berlekamp modular factoring algorithm, multivariate Hensel lifting algorithm and, eventually, different modular selection strategies for integer coefficients factoring. Also the univariate Kronecker algorithm based on the factorization in coefficient rings and interpolation will eventually be implemented. The presented implementation is only partially generic, as it is implemented for concrete coefficient rings (modular integers, integers and rational numbers). Only the algebraic number factorization is generic for iterated algebraic number coefficients. To achieve a fully generic design (called categorical in Axiom) as presented by [15] we need for example factorization over rational function fields (quotients of multivariate polynomials) and deal with inseparable field extensions[1].

8 Real Roots

In this section we sketch the implementation for real root isolation and applications. The algorithms are based at Sturm sequences, special polynomial

---

[1] By 2010 this has been worked out and will be presented elsewhere.

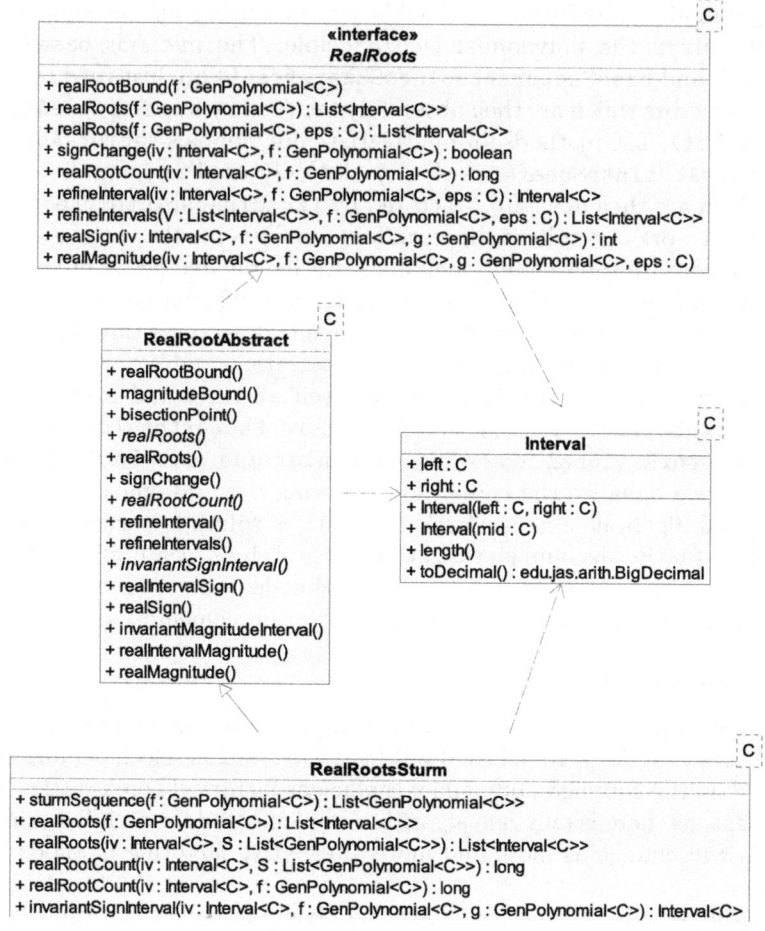

Fig. 18. Real Root classes and interfaces

remainder sequences, introduced in Sect. 4.1. The algorithms are implemented following Sect. 8.8 in [3] with the following changes. The last loop of algorithm ISOLATE with a hidden while loop is moved to algorithm ISOREC, which is implemented in method realRoots(v,S). The algorithm SQUEEZE is implemented in method realMagnitude(). Method realSign() was not in Sect. 8.8, but is crucial for the class RealAlgebraicNumber, as will be seen later. For mathematical background, different algorithms and references see Sect. 2.1.6 in [20] and [12].

The class design follows the standard pattern of interface, abstract and concrete classes, see Fig. 18. The interface RealRoots defines the visible API of the package. It defines the methods realRoots() to compute isolating intervals for the real roots and with given precision computes isolating intervals with length smaller than the requested precision. The method realRootCount(v,f) computes the number of real roots of the polynomial f in the interval v. Method realSign(v,f,g) with parameters, isolating interval v for a real root of the polynomial f computes the real sign $sign(g(\alpha))$ of polynomial g as element of the field $K[x]/f = K(\alpha)$. The method realMagnitude(v,f,g,eps) computes the magnitude $|g(\alpha)|$ of polynomial g as element of the field $K[x]/f = K(\alpha)$ up to a desired precision eps. Intervals are implemented with the container class Interval, which has a method toDecimal() to compute a floating point approximation of the interval mid-point.

The abstract class RealRootsAbstract implements methods which are independent of Sturm sequences. Finally class RealRootsSturm implements the remaining methods which require Sturm sequences. These are the methods realRootCount(), the realRoots() method without precision argument and the method invariantSignInterval() which is used for the computation of the real sign. For these methods exist companion methods which take a Sturm sequence as further parameter. And there is the method sturmSequence() to compute a Sturm sequence for a given polynomial. All other methods, mainly dealing with interval refinement, can be implemented in the abstract class.

8.1 Real Algebraic Numbers

Using the real root computation we can now construct classes for real algebraic numbers. They are represented as polynomials g modulo a defining polynomial f for the algebraic field extension together with an isolating interval for a specific real root α of f, that is $g(\alpha) \in K(\alpha) = K[x]/f$. The implementation is contained in class RealAlgebraicNumber with factory class RealAlgebraicRing. The factory class contains an instance of the real root computation engine, which is used to refine intervals as required.

A first version of these classes tried to extend class AlgebraicNumber respectively factory AlgebraicNumberRing from package edu.jas.poly. However, with this design it was not possible to define polynomials with such coefficients, due to the problem of type erasure in generic interfaces (see also Sect. 3.2 and the already mentioned [33, 31]). The solution is to use an object variable holding the AlgebraicNumber, that is, we use an association instead of sub-classing.

This design allows then the definition of polynomials with real algebraic coefficients

GenPolynomial<RealAlgebraicNumber<BigRational>>.

Moreover, for such polynomials we can also use real root isolation algorithms and instantiate for example

RealRootsSturm<RealAlgebraicNumber<BigRational>>.

This is possible since we implemented method realSign() which is used in method signum() of the a real algebraic number.

A slight optimization is implemented to use rational interval end points, but real algebraic end points work also. There is a small problem with method toDecimal() in class Interval which must use case distinction on types and casts to determine the correct way to compute rational end points. This will be solved in future by a suitable interface which extends `RingElem` and defines a method `toRational()`. `BigRational` and `RealAlgebraicNumber` will the implement this interface.

Left to do is the implementation of algorithm `REALZEROS` which computes the real roots of a zero dimensional ideal. This requires the computation of the univariate polynomials of minimal degree contained in the ideal, and their factorization[2].

9   Conclusions

We have designed and implemented a first part of 'multiplicative ideal theory', namely the computation of multivariate polynomial greatest common divisors and multivariate polynomial factorization. The new package provides a factory `GCDFactory` for the selection of one of the several implementations for non experts. The correct selection is based on the coefficient type of the polynomials and during run-time on the question if the coefficient ring is a field. A parallel proxy `GCDProxy` for gcd computation runs different implementations in parallel and takes the result from the first finishing method. This greatly improves the run-time behavior of the gcd computations. The new package is also type-safe designed with Java's generic types. We exploited the gcd package in the `Quotient`, `Residue` and `Product` classes. This provided a new coefficient ring of rational functions for the polynomials and also new coefficient rings of residue class rings and product rings. With an efficient gcd implementation we are now able to compute Gröbner bases over those coefficient rings. For small Gröbner base computations the performance is equal to MAS and for bigger examples the computing time with JAS is better by a factor of two or three. For random polynomials the performance of gcd computations is comparable to Maple.

For the factorization of polynomials we provide first versions for coefficient rings of modular integers, integers, rational numbers and algebraic numbers. Absolute factorization is implemented for rational numbers, modular integers and algebraic extensions thereof. We will have to fill some missing links to provide a fully generic design as presented by [15][3]. Future topics to explore, include the improvement, completion and performance of the factorization of polynomials. Eventually we will further investigate a new recursive polynomial representation for gcd computation and factorization.

Real roots of univariate polynomials can be computed by Sturm sequences and counting sign variations on interval bounds to arbitrary precision. The implementation is generic and can be used for polynomials with rational number coefficients and, moreover for real algebraic coefficients. There are faster algorithms for real root isolation which will eventually be implemented in the future, see [12]. Moreover complex root isolation is a topic to be treated in the future.

---

[2] By 2010 this has been worked out and will be presented elsewhere.

[3] This will be presented elsewhere in 2010.

Acknowledgments

I thank Thomas Becker for discussions on the implementation of a polynomial template library and Raphael Jolly for the discussions on the generic type system suitable for a computer algebra system. With Samuel Kredel I had many discussions on implementation choices for algebraic algorithms compared to C++. Thanks also to Markus Aleksy and Hans-Guenther Kruse for encouraging and supporting this work. JAS itself has improved by requirements from Axel Kramer and Brandon Barker. This paper incorporates the valuable feedback during ADG workshop from its participants, in particular Dongming Wang and Thomas Sturm to name a few.

References

1. Arnold, K., Gosling, J., Holmes, D.: The Java Programming Language, 4th edn. Addison-Wesley, Reading (2005)
2. Becker, M.Y.: Symbolic Integration in Java. PhD thesis, Trinity College, University of Cambridge (2001)
3. Becker, T., Weispfenning, V.: Gröbner Bases - A Computational Approach to Commutative Algebra. Graduate Texts in Mathematics. Springer, Heidelberg (1993)
4. Bernardin, L., Char, B., Kaltofen, E.: Symbolic computation in Java: an appraisement. In: Dooley, S. (ed.) Proc. ISSAC 1999, pp. 237–244. ACM Press, New York (1999)
5. Bini, D.A., Mourrain, B.: Polynomial test suite from PoSSo and Frisco projects. Technical report (1998), http://www-sop.inria.fr/saga/POL/ (accessed November 2009)
6. Böge, W., Gebauer, R., Kredel, H.: Some examples for solving systems of algebraic equations by calculating Groebner bases. J. Symb. Comput. 2(1), 83–98 (1986)
7. Bronstein, M.: Sigma^{it} - a strongly-typed embeddable computer algebra library. In: Limongelli, C., Calmet, J. (eds.) DISCO 1996. LNCS, vol. 1128, pp. 22–33. Springer, Heidelberg (1996)
8. Buchmann, J., Pfahler, T.: LiDIA. In: Computer Algebra Handbook, pp. 403–408. Springer, Heidelberg (2003)
9. Chen, X., Wang, D.: Towards an electronic geometry textbook. In: Botana, F., Recio, T. (eds.) ADG 2006. LNCS (LNAI), vol. 4869, pp. 1–23. Springer, Heidelberg (2007)
10. Cohen, H.: A Course in Computational Algebraic Number Theory. Springer, Heidelberg (1996)
11. Collins, G.E., Loos, R.G.: ALDES and SAC-2 now available. ACM SIGSAM Bull. 12(2), 19 (1982)
12. Collins, G.E., Loos, R.G.: Real Zeros of Polynomials. In: Buchberger, Collins, Loos (eds.) Computing Supplement: Computer Algebra, pp. 83–94. Springer, Heidelberg (1982)
13. Conrad, M.: The Java class package com.perisic.ring. Technical report (2002-2004), http://ring.perisic.com/ (accessed November 2009)
14. Dautelle, J.-M.: JScience: Java tools and libraries for the advancement of science. Technical report (2005-2007), http://www.jscience.org/ (accessed November 2009)
15. Davenport, H.J., Gianni, P., Trager, B.M.: Scratchpad's view of algebra II: A categorical view of factorization. In: Proc. ISSAC 1991, Bonn, pp. 32–38. ACM Press, New York (1991)
16. Gamma, E., Helm, R., Johnson, R., Vlissides, J.: Design Patterns. Addison-Wesley, Reading (1995); Deutsch: Entwurfsmuster. Addison-Wesley Reading (1996)

17. Geddes, K.O., Czapor, S.R., Labahn, G.: Algorithms for Computer Algebra. Kluwer, Dordrecht (1993)
18. Georges, A., Buytaert, D., Eeckhout, L.: Statistically rigorous Java performance evaluation. In: ACM Conference on Object-Oriented Programming Systems, Languages, and Applications, pp. 57–76. University of Montreal, Quebec (2007)
19. Gräbe, H.-G.: The SymbolicData project. Technical report (2000-2006), http://www.symbolicdata.org (accessed November 2009)
20. Grabmaier, J., Kaltofen, E., Weispfenning, V. (eds.): Computer Algebra Handbook. Springer, Heidelberg (2003)
21. Greuel, G.-M., Pfister, G., Schönemann, H.: Singular - A Computer Algebra System for Polynomial Computations. In: Computer Algebra Handbook, pp. 445–450. Springer, Heidelberg (2003)
22. Gruntz, D., Weck, W.: A Generic Computer Algebra Library in Oberon. Manuscript available via Citeseer (1994)
23. Hohenwarter, M., Fuchs, K.: Combination of dynamic geometry, algebra and calculus in the software system GeoGebra. In: Computer Algebra Systems and Dynamic Geometry Systems in Mathematics Teaching Conference, Pecs, Hungary (2004)
24. Jenks, R., Sutor, R. (eds.): axiom The Scientific Computation System. Springer, Heidelberg (1992)
25. Jolly, R.: jscl-meditor - Java symbolic computing library and mathematical editor. Technical report (2003), http://jscl-meditor.sourceforge.net/ (accessed November 2009)
26. Jolly, R., Kredel, H.: How to turn a scripting language into a domain specific language for computer algebra. Technical report (2008), http://arXiv.org/abs/0811.1061
27. Kaltofen, E.: Factorization of Polynomials. In: Buchberger, Collins, Loos (eds.) Computing Supplement: Computer Algebra, pp. 95–113. Springer, Heidelberg (1982)
28. Knuth, D.E.: The Art of Computer Programming - Volume 2, Seminumerical Algorithms. Addison-Wesley, Reading (1981)
29. Kredel, H.: On the Design of a Java Computer Algebra System. In: Proc. PPPJ 2006, pp. 143–152. University of Mannheim (2006)
30. Kredel, H.: Evaluation of a Java Computer Algebra System. In: Proceedings ASCM 2007, pp. 59–62. National University of Singapore (2007)
31. Kredel, H.: Evaluation of a Java computer algebra system. In: Kapur, D. (ed.) ASCM 2007. LNCS (LNAI), vol. 5081, pp. 121–138. Springer, Heidelberg (2008)
32. Kredel, H.: Multivariate greatest common divisors in the Java Computer Algebra System. In: Proc. Automated Deduction in Geometry (ADG), pp. 41–61. East China Normal University, Shanghai (2008)
33. Kredel, H.: On a Java Computer Algebra System, its performance and applications. Science of Computer Programming 70(2-3), 185–207 (2008)
34. Kredel, H.: Comprehensive Gröbner bases in a Java Computer Algebra System. In: Proceedings ASCM 2009, pp. 77–90. Kyushu University, Fukuoka (2009)
35. Kredel, H.: Distributed parallel Gröbner bases computation. In: Proc. Workshop on Engineering Complex Distributed Systems (ECDS) at CISIS 2009, Pages on CD–ROM, University of Fukuoka, Japan (2009)
36. Kredel, H.: Distributed hybrid Gröbner bases computation. In: Proc. Workshop on Engineering Complex Distributed Systems (ECDS) at CISIS 2010, University of Krakow, Poland (2010) (page to appear)

37. Kredel, H.: The Java algebra system (JAS), Technical report (2000), http://krum.rz.uni-mannheim.de/jas/
38. Kredel, H., Pesch, M.: MAS: The Modula-2 Algebra System. In: Computer Algebra Handbook, pp. 421–428. Springer, Heidelberg (2003)
39. Lewis, R.H., Wester, M.: Comparison of polynomial-oriented computer algebra systems. SIGSAM Bull. 33(4), 5–13 (1999)
40. Liang, T., Wang, D.: Towards a geometric-object-oriented language. In: Hong, H. (ed.) Proc. Automated Deduction in Geometry 2004, Florida, USA (2004)
41. Loos, R.: Generalized Polynomial Remainder Sequences. In: Buchberger, Collins, Loos (eds.) Computing Supplement: Computer Algebra, pp. 115–138. Springer, Heidelberg (1982)
42. Musser, D.R., Schupp, S., Loos, R.: Requirement oriented programming. In: Jazayeri, M., Musser, D.R., Loos, R.G.K. (eds.) Dagstuhl Seminar 1998. LNCS, vol. 1766, pp. 12–24. Springer, Heidelberg (2000)
43. Niculescu, V.: A design proposal for an object oriented algebraic library. Technical report, Studia Universitatis "Babes-Bolyai" (2003)
44. Niculescu, V.: OOLACA: an object oriented library for abstract and computational algebra. In: OOPSLA Companion, pp. 160–161. ACM, New York (2004)
45. Noro, M., Takeshima, T.: Risa/Asir - a computer algebra system. In: Proc. ISSAC 1992, pp. 387–396. ACM Press, New York (1992)
46. Platzer, A.: The Orbital library. Technical report, University of Karlsruhe (2005), http://www.functologic.com/
47. Sato, Y., Suzuki, A.: Gröbner bases in polynomial rings over von Neumann regular rings – their applications. In: Proceedings ASCM 2000. Lecture Notes Series on Computing, vol. 8, pp. 59–62. World Scientific Publications, Singapore (2000)
48. Schreiner, W., Hong, H.: PACLIB — a system for parallel algebraic computation on shared memory computers. In: Alnuweiri, H.M. (ed.) Parallel Systems Fair at the Seventh International Parallel Processing Symposium, IPPS 1993, Newport Beach, CA, April 14, pp. 56–61 (1993)
49. Schupp, S., Loos, R.: SuchThat - generic programming works. In: Jazayeri, M., Musser, D.R., Loos, R.G.K. (eds.) Dagstuhl Seminar 1998. LNCS, vol. 1766, pp. 133–145. Springer, Heidelberg (2000)
50. Stein, W.: SAGE Mathematics Software (Version 2.7). The SAGE Group (2007), http://www.sagemath.org (accessed November 2009)
51. Sun Microsystems, Inc. The Java development kit. Technical report (1994-2009), http://java.sun.com/ (accessed November 2009)
52. Watt, S.: Aldor. In: Computer Algebra Handbook, pp. 265–270. Springer, Heidelberg (2003)
53. Weispfenning, V.: Gröbner bases for polynomial ideals over commutative regular rings. In: Davenport, J.H. (ed.) ISSAC 1987 and EUROCAL 1987. LNCS, vol. 378, pp. 336–347. Springer, Heidelberg (1989)
54. Weispfenning, V.: Comprehensive Gröbner bases and regular rings. J. Symb. Comput. 41, 285–296 (2006)
55. Whelan, C., Duffy, A., Burnett, A., Dowling, T.: A Java API for polynomial arithmetic. In: Proc. PPPJ 2003, pp. 139–144. Computer Science Press, New York (2003)
56. Ye, Z., Chou, S.-C., Gao, X.-S.: An introduction to Java geometry expert (JGEX). In: Proc. Automated Deduction in Geometry (ADG), pp. 79–85. East China Normal University, Shanghai (2008)

Automatic Verification of the Adequacy of Models for Families of Geometric Objects

Aless Lasaruk[1] and Thomas Sturm[2]

[1] FORWISS, Universität Passau, 94030 Passau, Germany
`lasaruk@uni-passau.de`
[2] Departamento de Matemáticas, Estadística y Computación, Facultad de Ciencias,
Universidad de Cantabria, 39071 Santander, Spain
`sturmt@unican.es`

Abstract. We consider parametric families of semi-algebraic geometric objects, each implicitly defined by a first-order formula. Given an unambiguous description of such an object family and an intended alternative description we automatically construct a first-order formula which is true if and only if our alternative description uniquely describes geometric objects of the reference description. We can decide this formula by applying real quantifier elimination. In the positive case we furthermore derive the defining first-order formulas corresponding to our new description. In the negative case we can produce sample points establishing a counterexample for the uniqueness. We demonstrate our method by automatically proving uniqueness theorems for characterizations of several geometric primitives and simple complex objects. Finally, we focus on tori, characterizations of which can be applied in spline approximation theory with toric segments. Although we cannot yet practically solve the fundamental open questions in this area within reasonable time and space, we demonstrate that they can be formulated in our framework. In addition this points at an interesting and practically relevant challenge problem for automated deduction in geometry in general.

Keywords: Real Geometry, Unique Representation, Automated Proving, Real Quantifier Elimination.

1 Introduction

Motivated by questions from approximation theory with toric splines [10, 16, 11, 8, 13, 15], we are interested in automatically checking the adequacy of intended models for given families of geometric objects. Such a model is a family of real tuples, called model parameters, collecting coordinates of certain data, e.g. points, radii, normals, describing an object of the family. In an adequate model each model parameter describes a unique object of the model. In an unambiguous model, vice versa, each object of the model is described by only one model parameter.

Our approach is the following: For a given family of geometric objects we pick an unambiguous reference model with respect to which we specify our intended

T. Sturm and C. Zengler (Eds.): ADG 2008, LNAI 6301, pp. 116–140, 2011.

model. The intended semantics of our intended model is described by a first-order formula over the reals with ordering, which relates variables over the reference model with variables over the intended model. Similarly, the tuples admissible as model parameters for the reference model and the intended models are described by respective first-order formulas. From these formulas we automatically generate another first-order sentence, which establishes a hypothesis on the adequacy of our intended model. In a final—and computationally hardest—step, we prove or disprove our hypothesis using real quantifier elimination methods.

For a survey of implemented real quantifier elimination methods see e.g. [7]. We are also going to use some established generalizations of quantifier elimination. On the one hand, there is *generic quantifier elimination,* which has been originally introduced in the context of geometric theorem proving [17, 6]: Generic quantifier elimination possibly introduces during the elimination a set A of assumptions, which are exclusively negated equations in non-quantified variables. The assumptions are chosen in such a way that they support the elimination process by avoiding case distinctions on the vanishing of certain terms. For input φ and quantifier-free equivalent φ', we then have the semantics $\bigwedge A \longrightarrow (\varphi \longleftrightarrow \varphi')$. It is well-known that in the context of geometry the assumptions made by generic quantifier elimination are in general well-interpretable as non-degeneracy conditions [17, 6, 19, 18]. On the other hand there is *extended quantifier elimination,* which was originally motivated by elimination-based symbolic optimization methods, where it had been applied to existential formulas [23]. Here we are going to apply it, in contrast, to universal formulas. For such formulas it yields in case of unsatisfiability in addition to the quantifier-free equivalent "false" a set of sample values for the universally quantified variables that serve as a witness by *not* satisfying the quantifier-free part of the input formula. For a survey of both generic and extended quantifier elimination see e.g. [17].

On the software side we use the symbolic logic software REDLOG by the second author et al. [4], which is part of the computer algebra system REDUCE. REDUCE has recently been turned into an open-source project so that the entire system is freely available on the web[1]. In addition, the REDLOG homepage[2] features comprehensive documentation, references, and an online database containing computation examples. REDLOG provides implementations of several real quantifier elimination methods [7]. For our purposes here, we mostly use virtual substitution methods [9, 22] in combination with with sophisticated simplification techniques [5]. This is supplemented with partial cylindrical algebraic decomposition (CAD), which REDLOG uses by default as a fallback option for remaining quantifiers when virtual substitutions runs into degree violations. Whenever CAD is involved, we will explicitly point at this.

The original contributions of the present paper include the following:

1. For families of geometric objects we automatically generate a first-order formula that can be used to automatically decide via real quantifier elimination

---

[1] http://reduce-algebra.sourceforge.net/

[2] http://www.redlog.eu/

whether or not a given alternative description of the object family is suitable to represent these geometric objects uniquely.

2. In the positive case, we automatically generate by real quantifier elimination a quantifier-free description of the characteristic function with respect to the new representation.
3. In the negative case, we obtain by extended quantifier elimination a geometric configuration that establishes a counterexample for the uniqueness.
4. We rigorously simplify the resulting first-order formulas by transforming the objects of the reference model into "general position."
5. We make precise the asymptotic worst-case complexity of our approach.
6. We apply our framework to several non-trivial examples in real 2-space and real 3-space thus demonstrating its applicability. These examples include in particular a 2-dimensional variant of our following challenge problem on tori.
7. We propose a new challenging benchmark example for real quantifier elimination as well as for automated deduction in geometry in general: *Find necessary and sufficient conditions on finitely many points on a torus and normals in these points such that they uniquely describe a torus.* This we cannot solve ourselves yet but we provide various reductions of this problem, which we consider of general interest to the community.

The plan of our paper is as follows: In Sect. 2 we illustrate the basic idea of our paper by means of a simple example. In Sect. 3 we recall basic definitions of geometric objects and give a rigorous description of our technique. In Sect. 4 we make precise the asymptotic worst case complexity of our method. Sect. 5 then turns to concrete examples for geometric primitives in two and three dimensions. Sect. 6 describes work in progress: We give an overview on the current research on the approximation of parallel surfaces in particular by toric splines. We show that important questions on the quality of such approximations can be reduced to questions that are formally tractable within our framework described here. In Subsect. 6.1 we fully automatically treat the corresponding questions in 2-space, where they are nontrivial as well. In Sect. 7 we finally summarize and evaluate our results.

2   An Outline of Our Method

As an introductory example, we would like to check whether or not a sphere in the plane can be uniquely described by a point and a normal in this point on the sphere. In other words, are there two different spheres with a common point and normal in this point?

A sphere with center (c_x, c_y) and radius $r > 0$ is implicitly described by the equation

$$(x - c_x)^2 + (y - c_y)^2 = r^2.$$

Similarly, a point (p_x, p_y) lies on the sphere and has the normal (n_x, n_y) on the sphere if and only if

$$(p_x - c_x)^2 + (p_y - c_y)^2 = r^2 \quad \text{and} \quad -(p_x - c_x)n_y + (p_y - c_y)n_x = 0.$$

As usual, we require the normal vector to have length 1. Using this, our original question reduces to the question whether the following holds: For any choice of all occurring variables, whenever all of

$$r > 0, \qquad\qquad\qquad r_0 > 0,$$
$$(p_x - c_x)^2 + (p_y - c_y)^2 = r^2, \qquad (p_x - c_{x0})^2 + (p_y - c_{y0})^2 = r_0^2,$$
$$-(p_x - c_x)n_y + (p_y - c_y)n_x = 0, \qquad -(p_x - c_{x0})n_y + (p_y - c_{y0})n_x = 0,$$
$$n_x^2 + n_y^2 = 1$$

hold, then $(c_x, c_y) = (c_{x0}, c_{y0})$ and $r = r_0$. This straightforwardly describes that any two admissible spheres are equal.

Since one of the spheres can be freely chosen, it is "obviously" sufficient to consider $(c_x, c_y) = (0, 0)$ and $r = 1$. Substitution yields the following simplified formulation, where we formalize our implicit logical structure and universal quantification above:

$$\forall p_x \forall p_y \forall n_y \forall n_x \forall c_{x0} \forall c_{y0} \forall r_0 \, (p_x^2 + p_y^2 = 1 \wedge -p_x n_y + p_y n_x = 0 \wedge r_0 > 0 \wedge$$
$$(p_x - c_{x0})^2 + (p_y - c_{y0})^2 = r_0^2 \wedge -(p_x - c_{x0})n_y + (p_y - c_{y0})n_x = 0 \wedge$$
$$n_x^2 + n_y^2 = 1 \longrightarrow c_{x0} = 0 \wedge c_{y0} = 0 \wedge r_0 = 1).$$

Of course, the answer is negative: Consider, e.g., $(c_{x0}, c_{y0}) = (-1, 0)$, $r_0 = 2$, $(p_x, p_y) = (1, 0)$, and $(n_x, n_y) = (-1, 0)$. We are going to revisit this example within our formal framework as Example 3 in the following section. There the situation is pictured in Fig. 2.

The above procedure of modelling the problem and the simplifications used there are quite what any mathematician would intuitively do. Our goal is to formalize this apparently trivial procedure in order to *fully automatize* it.

Notice that first-order formulations like our final one given above can be directly solved by real quantifier elimination. Hence, combining our approach introduced here with real quantifier elimination, we arrive at a fully automatic procedure where quantifier elimination serves as a black box so that potential users need not care about first-order logic at all.

3   A Formal Description of Our Method

We consider real n-space $V = \mathbb{R}^n$. We consider furthermore a set $\Theta \subseteq \mathbb{R}^k$ and a function $\chi : V \times \Theta \to \{\text{true}, \text{false}\}$. For each $\theta \in \Theta$ we define

$$G_\chi(\theta) = \{ v \in V \mid \chi(v, \theta) \} \subseteq V.$$

For technical reasons $G_\chi(\theta)$ may be the trivial object $\emptyset$. This way we obtain a family $G_\chi = \{G_\chi(\theta)\}_{\theta \in \Theta}$ of geometric objects over V. We call (Θ, χ) a *model* of the set of objects $G_\chi[\Theta] = \{ G_\chi(\theta) \in \wp(V) \mid \theta \in \Theta \}$. We call each $\theta \in \Theta$ a *model parameter*, and we call χ the *characteristic function* of (Θ, χ). Notice that χ naturally corresponds to a relation on $V \times \Theta$.

For our framework discussed in this paper, we are going to consider first-order formulas over the reals with ordering. For simplicity, we are generally going to refer to such formulas simply as *formula*. We generally assume that Θ can be described by a formula $\tau(t_1, \ldots, t_k)$ and that χ is given as formula $\chi(x_1, \ldots, x_n, t_1, \ldots, t_k)$.

A model is called *unambiguous* if for each object $G \neq \emptyset$ in $G_\chi[\Theta]$ there is exactly one model parameter $\theta \in \Theta$ such that $G = G_\chi(\theta)$.

Example 1 (Spheres in 2-space). Consider $V = \mathbb{R}^2$ and the family of spheres in V. We choose the model (Θ, χ) with $\Theta = \mathbb{R}^2 \times \mathbb{R}^>$ given by $\tau(c_x, c_y, r) \equiv r > 0$ and

$$\chi(x, y, c_x, c_y, r) \equiv (x - c_x)^2 + (y - c_y)^2 = r^2.$$

That is, each model parameter $\theta \in \Theta$ describes the center and the radius of a corresponding sphere. This model is unambiguous. An ambiguous variant can be obtained by generalizing to $\Theta = \mathbb{R}^2 \times \mathbb{R} \setminus \{0\}$; then the radius r can be alternatively represented by $-r$. □

Consider a model (Θ, χ) with $\Theta \subseteq \mathbb{R}^k$, which is going to serve as a reference model. We choose a set $\Theta' \subseteq \mathbb{R}^l$, which we would like to establish another model for the same family of geometric objects as (Θ, χ). An *intended semantics* of Θ' with respect to (Θ, χ) is a formula $\psi(t, t')$ where t and t' are vectors of variables ranging over Θ and Θ', respectively, such that for each $\theta \in \Theta$ there exists $\theta' \in \Theta'$ such that $\psi(\theta, \theta')$ holds. That is, ψ assigns to each $\theta \in \Theta$ at least one $\theta' \in \Theta'$. Notice that given formulas $\tau(t)$ and $\tau'(t')$ for Θ and Θ', respectively, this defining condition can be automatically checked by real quantifier elimination applied to $\forall t(\tau \longrightarrow \exists t'(\tau' \wedge \psi))$.

Example 2 (Spheres in 2-space, continued). We continue the previous example. We pick $\Theta' = \mathbb{R}^2 \times S_2$, where $S_2 \subseteq \mathbb{R}^2$ is the unit sphere, described by $\tau'(p_x, p_y, n_x, n_y) \equiv n_x^2 + n_y^2 = 1$. Our intended semantics is given by

$$\psi(c_x, c_y, r, p_x, p_y, n_x, n_y) \equiv (p_x - c_x)^2 + (p_y - c_y)^2 = r^2 \wedge$$
$$-(p_x - c_x)n_y + (p_y - c_y)n_x = 0.$$

This semantics straightforwardly describes the intention that (p_x, p_y) is a point on the modelled sphere and (n_x, n_y) is a normal vector on the sphere at that point. □

Let $\Sigma \subseteq \Theta$. An intended semantics ψ of Θ' is *adequate for* (Σ, χ) *with respect to* (Θ, χ) if the following holds:

(i) There exists χ^* such that (Θ', χ^*) is a model for some G_{χ^*} comprising the same nontrivial objects as $G_\chi|_\Sigma$.

(ii) If there are $\sigma \in \Sigma$, $\theta \in \Theta$, and $\theta' \in \Theta'$ such that both $\psi(\sigma, \theta')$ and $\psi(\theta, \theta')$ hold, then $G_\chi(\sigma) = G_\chi(\theta) = G_{\chi^*}(\theta')$.

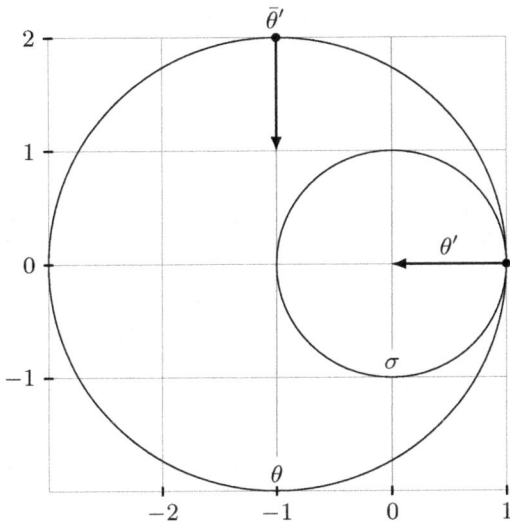

Fig. 1. An illustration of the definition of adequacy

Since this definition is quite subtle, we give an example, which is illustrated in Fig. 1: Choose (Θ, χ) to model the family of spheres, and choose $\Sigma \subseteq \Theta$ as the singleton containing the unit sphere. We have $\Sigma = \{\sigma\}$ and $\theta \in \Theta \setminus \Sigma$. Consider an intended semantics ψ that assigns to a sphere an element of Θ', which is a point on that sphere and a normal vector in that point, e.g., $\psi(\sigma, \theta')$ and $\psi(\theta, \theta')$, and $\psi(\theta, \bar{\theta}')$. Although $\theta \notin \Sigma$, this ψ is not adequate for Σ with respect to (Θ, χ) because $G_\chi(\sigma) \neq G_\chi(\theta)$. On the other hand, $\psi(\theta, \bar{\theta}')$ cannot possibly turn ψ inadequate, because according to our intuitive specification above ψ cannot possibly assign $\bar{\theta}'$ also to the unit sphere σ, which is the only object modelled by Σ.

In the important special case that $\Sigma = \Theta$, we simply say that the intended semantics ψ of Θ' is *adequate for* (Θ, χ). Then for $\chi' := \chi^*$ we have that (Θ', χ') comprises the same nontrivial objects as G_χ, and condition (ii) simplifies as follows: If there are $\theta \in \Theta$ and $\theta' \in \Theta'$ such that $\psi(\theta, \theta')$, then $G_\chi(\theta) = G_{\chi'}(\theta')$. Recall that by the definition of intended semantics, there is at least one θ' for each θ satisfying $\psi(\theta, \theta')$, and for unambiguous (Θ, χ) there is at most one $\theta \in \Theta$ for each θ' satisfying $\psi(\theta, \theta')$.

Example 3 (Spheres in 2-space, continued). Consider $\theta = (-1, 0, 2) \in \Theta$ and $\theta' = (1, 0, -1, 0) \in \Theta'$. Consider furthermore the sphere $\theta^* = (0, 0, 1) \in \Theta$. The situation is depicted in Fig. 2. Obviously, $G_\chi(\theta) \neq G_\chi(\theta^*)$. On the other hand, one easily verifies that both $\psi(\theta, \theta')$ and $\psi(\theta^*, \theta')$ hold. Assume for a contradiction that there is χ' as required by the definition of adequacy. It then follows that

$$G_{\chi'}(\theta') = G_\chi(\theta) \neq G_\chi(\theta^*) = G_{\chi'}(\theta').$$

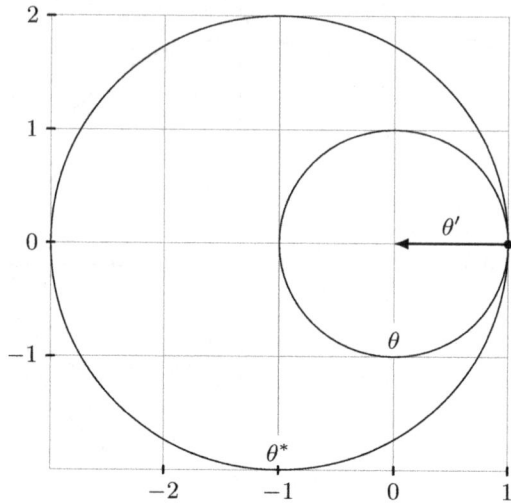

Fig. 2. An illustration of the geometry in Example 3

Thus our specification of one point with a normal vector is not adequate for spheres.   □

In Example 3 we have given a manual proof for the fact that spheres in real 2-space cannot adequately be described by a single point in combination with a normal vector on the sphere at this point. The remainder of this section is devoted to automatizing such proofs within our framework. For our approach it is necessary to generally start with unambiguous reference models. For our intended models, in contrast, we are not interested in unambiguity. Recall from the definitions that even in an ambiguous model (Θ', χ') each model parameter still describes a unique geometric object because χ' is a function. It is just not necessarily the only model parameter describing this object.

Theorem 4. *Let (Θ, χ), where $\Theta \subseteq \mathbb{R}^k$ is described by $\tau(t)$, be an unambiguous model for a family G_χ of geometric objects in $\mathbb{R}^n$. Consider $\Theta' \subseteq \mathbb{R}^l$ described by $\tau'(t')$ and an intended semantics $\psi(t, t')$ of Θ' with respect to (Θ, χ). Then the following holds:*

(i) The following first-order formula over the reals is equivalent to the adequacy of ψ for G_χ:

$$\Phi(\psi, t, \tau, \tau') \equiv \forall t' \forall t \forall t_0 (\tau \wedge \tau[t_0/t] \wedge \tau' \longrightarrow \psi \wedge \psi[t_0/t] \longrightarrow t = t_0).$$

(ii) In case of adequacy χ' is given by $\chi' \equiv \exists t (\tau \wedge \chi \wedge \psi)$.   □

Using real quantifier elimination one can automatically decide the condition Φ in (i) and in the positive case obtain a quantifier-free description of χ' in (ii).

Furthermore, in (i) we may drop from Φ some of the universal quantifiers and use generic quantifier elimination, which speeds up the elimination process. As a rule one would drop quantifiers from the outermost block $\forall t'$, which refer to the intended model Θ'. Then generic quantifier elimination possibly yields non-degeneracy conditions on Θ'. In the negative case extended quantifier elimination can produce a counterexample proving that ψ is not adequate. Recall that generally extended and generic quantifier elimination can be combined.

We are going to model and automatically prove our example on spheres in real 2-space, which we had discussed in Example 1–3.

Example 5 (Spheres in 2-space, continued). Recall that we have got

$$\tau(c_x, c_y, r) \equiv r > 0,$$
$$\tau'(p_x, p_y, n_x, n_y) \equiv n_x^2 + n_y^2 = 1,$$
$$\psi(c_x, c_y, r, p_x, p_y, n_x, n_y) \equiv (p_x - c_x)^2 + (p_y - c_y)^2 = r^2 \wedge$$
$$-(p_x - c_x)n_y + (p_y - c_y)n_x = 0.$$

Using a procedure implemented in REDUCE for this purpose, we automatically generate

$$\Phi(\psi, \{c_x, c_y, r\}, \tau, \tau') \equiv$$
$$\forall p_x \forall p_y \forall n_y \forall n_x \forall c_x \forall c_y \forall r \forall c_{x0} \forall c_{y0} \forall r_0 (r > 0 \wedge r_0 > 0 \wedge n_x^2 + n_y^2 - 1 = 0$$
$$\longrightarrow (p_x - c_x)^2 + (p_y - c_y)^2 = r^2 \wedge -(p_x - c_x)n_y + (p_y - c_y)n_x = 0 \wedge$$
$$(p_x - c_{x0})^2 + (p_y - c_{y0})^2 = r_0^2 \wedge -(p_x - c_{x0})n_y + (p_y - c_{y0})n_x = 0$$
$$\longrightarrow c_x - c_{x0} = 0 \wedge c_y - c_{y0} = 0 \wedge r - r_0 = 0).$$

We are going to continue the discussion of this Φ as a part of Example 10 in the next section.   □

We are now going to discuss how the formulas produced by the application of Theorem 4 can be simplified by bringing the reference model into *general position*.

Let (Θ, χ) be an unambiguous model of some family G_χ of geometric objects not containing $\emptyset$. A group F of functions *operating on* G_χ is a group of functions operating on V with the following closure property: For each $f \in F$ and $\theta \in \Theta$ there is $\theta_0 \in \Theta$ such that $f(G_\chi(\theta)) = G_\chi(\theta_0)$. We obtain from F an equivalence relation $\sim_F$ on Θ by defining

$$\theta \sim_F \theta_0 \iff f(G_\chi(\theta)) = G_\chi(\theta_0) \text{ for some } f \in F.$$

Consider now Θ' and some intended semantics ψ for Θ' with respect to (Θ, χ). We call ψ *compatible* with F if for each $f : V \to V \in F$ there is $\bar{f} : \Theta' \to \Theta'$ such that for all $\theta, \theta_0 \in \Theta$ and $\theta' \in \Theta'$ with $f(G_\chi(\theta)) = G_\chi(\theta_0)$, we have

$$\psi(\theta, \theta') \implies \psi(\theta_0, \bar{f}(\theta')).$$

Lemma 6. *Let (Θ, χ) be an unambiguous model of some family G_χ of geometric objects not containing $\emptyset$. Let F be a group of functions operating on G_χ. Consider the model (Θ_F, χ), where Θ_F is a set of representants of $\Theta/\sim_F$. Let ψ be an intended semantics of some Θ' with respect to (Θ, χ) that is compatible with F. Then the following two statements are equivalent:*

(i) ψ is adequate for (Θ_F, χ) with respect to (Θ, χ).
(ii) ψ is adequate for (Θ, χ).

Proof. Assume (i). We construct $\chi' : V \times \Theta' \to \{\text{true}, \text{false}\}$ satisfying the adequacy condition in (ii) by specifying $G_{\chi'}(\theta')$ for given $\theta' \in \Theta'$. If there is no $\theta \in \Theta$ such that $\psi(\theta, \theta')$ we set $G_{\chi'}(\theta') = \emptyset$. Else there is exactly one $\theta_f \in \Theta_F$ such that $\theta_f \sim_F \theta$ via $f \in F$, i.e., $f(G_\chi(\theta)) = G_\chi(\theta_f)$. We set

$$G_{\chi'}(\theta') = f^{-1}(G_\chi(\theta_f)) = G_\chi(\theta),$$

which is well-defined: For let $\theta^* \in \Theta$ with $\psi(\theta^*, \theta')$. There is $\theta_g \in \Theta$ with $f(G_\chi(\theta^*)) = G_\chi(\theta_g)$. Since ψ is compatible with F we have $\psi(\theta_f, \bar{f}(\theta'))$ and $\psi(\theta_g, \bar{f}(\theta'))$. Since ψ is adequate for (Θ_F, χ) with respect to (Θ, χ), there exists χ^* such that (Θ', χ^*) satisfies the definition of adequacy, and we have

$$G_\chi(\theta_f) = G_{\chi^*}(\bar{f}(\theta')) = G_\chi(\theta_g).$$

We apply to both sides f^{-1} and obtain

$$G_\chi(\theta) = f^{-1}(G_\chi(\theta_f)) = f^{-1}(G_\chi(\theta_g)) = G_\chi(\theta^*).$$

From the unambiguity of (Θ, χ) it follows that $\theta = \theta^*$. From the definition of intended semantics it follows that for each $\theta \in \Theta$ there is at least one $\theta' \in \Theta'$ such that $\psi(\theta, \theta')$ holds. From our construction it follows that $G_\chi(\theta) = G_{\chi'}(\theta')$. Hence our new model comprises the same nontrivial objects as (Θ, χ).

The converse implication from (ii) to (i) is straightforward. $\square$

Throughout this paper we focus on sets of points and normals for our intended models Θ'. Possible choices for the function group F to bring a model into general position are then isometries or more generally affine bijections $f : \mathbb{R}^n \to \mathbb{R}^n$ with $f(x) = \alpha A x + b$, where A is an orthogonal matrix, $b \in \mathbb{R}^n$, and $\alpha \in \mathbb{R} \setminus \{0\}$. The group of bijections of the above form is called the group of *similarities*. In the case of similarities a suitable choice for $\bar{f}$ in the definition of compatibility is given by

$$\bar{f}(p_1, \ldots, p_k, n_1, \ldots, n_l) = (f(p_1), \ldots, f(p_k), An_1, \ldots, An_l).$$

Notice that further ascending in the transformation group hierarchy by allowing weaker conditions on group members, e.g. by allowing arbitrary affine bijections, does not generally yield a compatible description because the properties of normals are not necessarily preserved by these functions.

Theorem 7. *Let (Θ, χ), where $\Theta \subseteq \mathbb{R}^k$ is described by $\tau(t)$, be an unambigu-ous model for a family G_χ of geometric objects not containing $\emptyset$. Let $\Theta' \subseteq \mathbb{R}^l$ described by $\tau'(t')$. Consider an intended semantics $\psi(t, t')$ of Θ' with respect to (Θ, χ) that is compatible with a group of functions operating on Θ, where $\Theta_0 \subseteq \Theta$ is a set of representants of $\Theta/\sim$. Let σ be a substitution on the vari-ables t in τ defined as follows: Check in Θ_0 for which coordinates there occurs only one value; σ substitutes these unique values for the corresponding variables $s \subseteq t$. Then denoting $\tilde{t} = t \setminus s$ the following first-order formula over the reals is equivalent to the adequacy of ψ for G_χ:*

$$\tilde{\Phi}(\psi, \tilde{t}, \tau, \tau', \sigma) \equiv \forall t' \forall \tilde{t} \forall t_0 (\tau\sigma \wedge \tau[t_0/t] \wedge \tau' \longrightarrow \psi\sigma \wedge \psi[t_0/t] \longrightarrow t\sigma = t_0).$$

Proof. This follows immediately from Lemma 6.   □

Notice that in the positive case χ' can still be computed as described in Theo-rem 4 (ii). Furthermore the natural assumption that the reference model does not contain the empty set can be verified by quantifier elimination applied to the sentence $\forall t (\tau \longrightarrow \exists x \chi)$. Exactly as discussed for Theorem 4 we can drop some quantifiers and use generic quantifier elimination. It is important to understand, however, that any condition obtained this way, even for variables of the intended model, refers to the description of objects from $\Theta/\sim$.

In the sequel we generally choose the function group F and the set of repre-sentants Θ_0 such that we maximize the number of parameters that are equal in all entries of Θ_0. We continue our example of spheres in 2-space showing how the formulation in Example 5 is simplified by Lemma 6 and Theorem 7.

Example 8 (Spheres in 2-space, continued). With respect to the group of sim-ilarities in $\mathbb{R}^2$ we can choose the set $\Theta_0 = \{(0, 0, 1)\}$ as a set of represen-tants for spheres in 2-space. The corresponding subset of spheres contains only one element: the unit sphere in the origin. Hence we obtain the substitution $\sigma = [0/x_c, 0/y_c, 1/r]$. According to Theorem 7 we automatically generate essen-tially the following formula:

$$\tilde{\Phi}(\psi, \{c_x, c_y, r\}, \tau, \tau') \equiv$$
$$\forall p_x \forall p_y \forall n_y \forall n_x \forall c_{x0} \forall c_{y0} \forall r_0 (1 > 0 \wedge r_0 > 0 \wedge n_x^2 + n_y^2 = 1$$
$$\longrightarrow p_x^2 + p_y^2 = 1 \wedge -p_x n_y + p_y n_x = 0 \wedge$$
$$(p_x - c_{x0})^2 + (p_y - c_{y0})^2 = r_0^2 \wedge -(p_x - c_{x0})n_y + (p_y - c_{y0})n_x = 0$$
$$\longrightarrow c_{x0} = 0 \wedge c_{y0} = 0 \wedge r_0 = 1).$$

This formula is simpler than the one obtained in Example 5 in the following sense: It does not contain the three variables c_x, c_y, r and also not the corre-sponding universal quantifiers. We are going to continue the discussion of this $\tilde{\Phi}$ in combination with Φ from Example 5 as Example 10 in the next section.   □

4   Complexity

The complexity of our approach is obviously dominated by the real quantifier elimination step. Real quantifier elimination is well-known to have in general double exponential time complexity in the worst case [2, 20]. In our approach, however, we exclusively use universal quantifiers. The absence of quantifier changes reduces the worst-case complexity to single exponential. More precisely it is single exponential in the number of universal quantifiers but only polynomial in all other natural complexity parameters, as e.g. number of atomic formulas, polynomial degrees, polynomial coefficient sizes. Extended quantifier elimination and generic quantifier elimination are not essentially harder than regular quantifier elimination. Notice that with our use of generic quantifier elimination we drop some quantifiers, which exponentially improves the complexity.

5   Example Computations for Geometric Primitives

To illustrate our framework, we are going to study in this section several families of geometric primitives. We are going to automatically prove or disprove statements concerning the number of points and normals needed to uniquely—though not necessarily unambiguously—specify these objects. We start in two dimensions, where we consider the families of lines and spheres. In three dimensions we consider the families of planes and spheres. The following Sect. 6 includes some computations with complex objects, viz. circle rings in 2-space and tori in 3-space.

All our computations have been carried out on a 2.8 GHz Intel Core 2 Duo using only one core within at most 1 GB of memory.

5.1   Lines and Spheres in 2-Space

A line in 2-space is uniquely defined by one point on the line and one normal in this point. Despite the simplicity of this problem, we give an automatic proof of this statement in order to summarize our framework once more:

Example 9 (Lines in 2-space). A natural unambiguous reference model for the family of lines in 2-space is given by (Θ, χ), where $\Theta = \{\, \theta \in \mathbb{R}^3 \mid \tau(\theta) \,\}$ with

$$\tau(a, b, c) \equiv a^2 + b^2 = 1 \wedge (a > 0 \vee (a = 0 \wedge b = 1)),$$

and the characteristic function is given by

$$\chi(x, y, a, b, c) \equiv ax + by + c = 0.$$

We choose $\Theta' = \{\, \theta' \in \mathbb{R}^4 \mid \tau'(\theta') \,\}$ with

$$\tau'(p_x, p_y, n_x, n_y) \equiv n_x^2 + n_y^2 = 1$$

and the following intended semantics that (p_x, p_y) is a point on the line and (n_x, n_y) is a normal in this point:

$$\psi(a, b, c, p_x, p_y, n_x, n_y) \equiv ap_x + bp_y + c = 0 \wedge -n_y a + n_x b = 0.$$

Regular quantifier elimination applied to the formula

$$\forall a \forall b \forall c(\tau \longrightarrow \exists p_x \exists p_y \exists n_x \exists n_y(\tau' \wedge \psi))$$

yields "true" in less than 10 ms, which confirms that ψ matches the definition of an intended semantics. By directly applying Theorem 4 we automatically obtain the formula

$$
\begin{aligned}
&\Phi(\psi, \{a, b, c\}, \tau, \tau') \equiv \\
&\quad \forall p_x \forall p_y \forall n_y \forall n_x \forall a_0 \forall b_0 \forall c_0 (a^2 + b^2 = 1 \wedge (a > 0 \vee (a = 0 \wedge b = 1)) \wedge \\
&\quad a_0^2 + b_0^2 = 1 \wedge (a_0 > 0 \vee (a_0 = 0 \wedge b_0 = 1)) \wedge n_x^2 + n_y^2 = 1 \\
&\quad \longrightarrow a p_x + b p_y + c = 0 \wedge -n_y a + n_x b = 0 \wedge \\
&\quad \quad a_0 p_x + b_0 p_y + c_0 = 0 \wedge -n_y a_0 + n_x b_0 = 0 \\
&\quad \longrightarrow a_0 = a \wedge b_0 = b \wedge c_0 = c)
\end{aligned}
$$

as a necessary and sufficient condition for the adequacy of ψ. For this formula REDLOG returns in 30 ms the expected result "true."

With respect to the group of similarities we can choose the set of representants $\Theta_0 = \{(0, 1, 0)\}$, which exclusively contains the representation of the line identical to the x-axis. Since all elements of Θ_0 trivially have the same entries, we have $\sigma = [0/a, 1/b, 0/c]$. Using now Theorem 7 we automatically obtain the following alternative formula:

$$
\begin{aligned}
&\tilde{\Phi}(\psi, \{a, b, c\}, \tau, \tau') \equiv \\
&\quad \forall p_x \forall p_y \forall n_y \forall n_x \forall a_0 \forall b_0 \forall c_0 (\\
&\quad a_0^2 + b_0^2 = 1 \wedge (a_0 > 0 \vee (a_0 = 0 \wedge b_0 = 1)) \wedge n_x^2 + n_y^2 = 1 \\
&\quad \longrightarrow p_y = 0 \wedge n_x = 0 \wedge a_0 p_x + b_0 p_y + c_0 = 0 \wedge -n_y a_0 + n_x b_0 = 0 \\
&\quad \longrightarrow a_0 = 0 \wedge b_0 = 1 \wedge c_0 = 0).
\end{aligned}
$$

For this $\tilde{\Phi}$ REDLOG returns the expected result "true" in less than 10 ms.

Since our model has turned out adequate, we may apply part (ii) of Theorem 4 to obtain the corresponding characteristic function $\chi' = \exists a \exists b \exists c(\tau \wedge \chi \wedge \psi)$. Within 50 ms we obtain a quantifier-free description containing 76 atomic formulas, and within another 50 ms we can prove by quantifier elimination that this obtained description is in fact equivalent to the obvious characteristic function $(x - p_x)n_x + (y - p_y)n_y = 0$.   □

We continue with the computation results for our spheres in 2-space, which we had discussed repeatedly as Examples 1–3, 5, and 8.

Example 10 (Spheres in 2-space, continued). For the formula Φ given in Example 5 regular quantifier elimination yields "false" within 20 ms. Extended quantifier elimination yields within 60 ms in addition the following sample point:

$$
\begin{aligned}
\{ & c_x = -1, \ c_{x0} = 0, \ c_y = 0, \ c_{y0} = 0, \\
& n_x = 1, \ n_y = 0, \ p_x = -2, \ p_y = 0, \ r = 1, \ r_0 = 2 \}.
\end{aligned}
$$

Notice that this result is as natural as our illustrative counterexample from Example 3. In fact it is that example mirrored at the line $y = -1/2$.

For the formula $\tilde{\Phi}$ given in Example 8 we obtain "false" in 30 ms. Here extended quantifier elimination yields an alternative sample point:

$$\{c_{x0} = -2,\ c_{y0} = 0,\ n_x = 1,\ n_y = 0,\ p_x = -1,\ p_y = 0,\ r_0 = 1\}. \quad \square$$

We now modify our example on spheres in 2-space by adding a second point to our intended semantics thus considering the following statement: *Two distinct points together with a normal in one of these points comprise an adequate model for spheres in 2-space.* It is not hard to see that this statement is true: Consider the line segment connecting the two given points. A line through the middle of this segment and orthogonal to it contains the center of the circle. From this construction it is easy to obtain all parameters of the circle.

Example 11 (Spheres in 2-space by two points and one normal). We use the same Θ and Θ_0 with $\sigma = [0/c_x, 0/c_y, 1/r]$ as in Example 8. Thus τ remains unmodified. For our intended model we switch to $\Theta' = \{\theta' \in \mathbb{R}^6 \mid \tau'(\theta')\}$, where we add to τ' the condition that our two considered points are distinct:

$$\tau'(p_{1x}, p_{1y}, n_x, n_y, p_{2x}, p_{2y}) \equiv n_x^2 + n_y^2 = 1 \wedge (p_{1x} \neq p_{1y} \vee p_{2x} \neq p_{2y}).$$

We consider the following straightforward intended semantics:

$$\begin{aligned}
\psi(c_x, c_y, r, p_{1x}, p_{1y}, n_{1x}, n_{1y}, p_{2x}, p_{2y}) \equiv\ & (p_{1x} - c_x)^2 + (p_{1y} - c_x)^2 = r^2 \wedge \\
& (p_{2x} - c_x)^2 + (p_{2y} - c_x)^2 = r^2 \wedge \\
& -n_{1y}(p_{1x} - c_x) + n_{1x}(p_{1y} - c_y) = 0.
\end{aligned}$$

This way we automatically obtain via Theorem 7 our first-order input formula $\tilde{\Phi}$. From this we drop the outermost universal quantifiers $\forall p_{1x} \forall p_{1y} \forall n_x \forall n_y \forall p_{2x} \forall p_{2y}$, which refer to the variables for our intended model Θ'. This yields $\hat{\Phi}$. We apply generic quantifier elimination and obtain after 20 ms a quantifier-free equivalent $\hat{\Phi}'$ with 10 atomic formulas subject to the following assumptions:

$$A = \{n_x p_{1x} - n_x p_{2x} + n_y p_{1y} - n_y p_{2y} \neq 0,\ n_x \neq 0\}.$$

Regular quantifier elimination proves within 4.6 s that $\bigwedge A \longrightarrow \hat{\Phi}'$, i.e., $\hat{\Phi}'$ is true on the assumptions in A. For this elimination REDLOG uses for the last quantifier $\forall n_x$ partial CAD as a fallback option. The first condition in A states that the line through p_1 and p_2 is not orthogonal to n. This obviously follows from that facts that $p_1 \neq p_2$ and that n is a normal at p_1. As for the second condition, regular quantifier elimination on $\underline{\forall}\hat{\Phi}[0/n_x]$ yields "true" in less than 10 ms. This shows that this condition is not relevant.

A bit surprisingly, applying generic quantifier elimination in the same style to Φ according to Theorem 4 yields the same assumptions A as above plus "true" as a quantifier-free equivalent in only 20 ms.

Regular quantifier elimination succeeds neither for Φ nor for $\tilde{\Phi}$ within reasonable time and space. $\quad \square$

In the previous example we have made the important observation that generic quantifier elimination by virtual substitution performs better on the more general formula Φ according to Theorem 4 than on $\tilde{\Phi}$ according to Theorem 7. This is something which requires to be carefully monitored and analyzed in the future. For now, we collect some ideas, which are suitable to explain that anomaly: First, the substitution σ might not substantially support the elimination here and become irrelevant by simplification after the elimination of the first few quantifiers. Second, σ destroys symmetries in the input formula, the presence of which might support the simplification process during elimination. Finally, due to the extra quantified variables in Φ there is more freedom of choice for a good elimination order of the universal quantifiers, which can be freely interchanged. Notice also, that the observed effect might be caused by the fact that there is an elimination order chosen that simply happens to be better; in that case it would not be significant but point once more at the well-known fact that quantifier elimination is generally sensitive to the chosen order of like quantifiers [3].

5.2   Planes and Spheres in 3-Space

We now turn to computation examples in 3-space. For this we simplify notation as follows: For $u = (u_x, u_y, u_z)$ and $v = (v_x, v_y, v_z)$, we denote by $\langle u \mid v \rangle$ the polynomial resulting from the application of the standard scalar product to u and v. Similarly, $\|u\|^2$, which equals $\langle u \mid u \rangle$, denotes the square of the Euclidean norm of u. We denote by $u \times v = 0$ the conjunction resulting from equating each component of the cross product $u \times v$ to zero. Furthermore we admit cross products $u \times v$ as arguments of scalar products, which obviously leads to polynomial conditions as well.

Example 12 (Planes in 3-space). An unambiguous model for planes in 3-space is given by (Θ, χ), where $\Theta = \{\, \theta \in \mathbb{R}^4 \mid \tau(\theta) \,\}$ with

$$\tau(a, b, c, d) \equiv \|(a, b, c)\|^2 = 1 \,\wedge$$
$$(a > 0 \vee (a = 0 \wedge b > 0) \vee (a = 0 \wedge b = 0 \wedge c > 0)),$$

and the characteristic function is given by

$$\chi(x, y, z, a, b, c, d) \equiv \langle (x, y, z) \mid (a, b, c) \rangle + d = 0.$$

We consider $\Theta' = \{\, \theta \in \mathbb{R}^6 \mid \tau'(\theta) \,\}$ with $\tau'(p, n) \equiv \|n\|^2 = 1$ for $p = (p_x, p_y, p_z)$ and $n = (n_x, n_y, n_z)$. Our intended semantics is given by

$$\psi(a, b, c, d, p, n) \equiv \langle p \mid (a, b, c) \rangle + d = 0 \wedge n \times (a, b, c) = 0,$$

i.e., p and n are one point and one normal, respectively. Applying regular quantifier elimination to the formula Φ automatically obtained according to Theorem 4 does not succeed within reasonable time and space.

For generating $\tilde{\Phi}$ according to Theorem 7 we choose with respect to the similarities the set of representatives

$$\Theta_0 = \{(0, 1, 0, 0)\} \quad \text{with} \quad \sigma = [0/a, 1/b, 0/c, 0/d].$$

We then obtain by regular quantifier elimination "true" in less than 10 ms. By means of generic quantifier elimination we obtain within less than 10 ms a quantifier-free description for $\chi' \equiv \exists a \exists b \exists c \exists d (\tau \wedge \chi \wedge \psi)$, viz.

$$n_x p_x - n_x x + n_y p_y - n_y y + n_z p_z - n_z z = 0,$$

i.e. $\langle n \mid p - (x, y, z) \rangle = 0$, subject to the condition $A = \{n_x \neq 0\}$. For this, we have not used the default settings of REDLOG but switched off `rlgenct`, which stands for *generate complex theory*. This way, generic quantifier elimination may exclusively make monomial assumptions. □

With the standard settings, i.e. `rlqegenct` on, we obtain in the previous example "false" as a quantifier-free equivalent for χ'. The conditions A then contain $\langle n \mid p - (x, y, z) \rangle \neq 0$ which exactly contradicts our model that n is normal on the plane. This makes the result meaningless. The observation that with orthogonality conditions generic quantifier elimination produces assumptions that are not non-degeneracy conditions has to our knowledge not been reported in the literature so far.

A sphere in 3-space is uniquely given by two points and two normals in these points. The center, and subsequently the radius, can be computed by intersecting the rays starting in the two given points in direction of the respective normals. In fact, a sphere is uniquely defined even by two points and only one normal in one of the points. This is less obvious; a non-automiatic proof can be obtained by a construction in analogy to the spheres example in 2-space.

Example 13 (Spheres in 3-space). An unambiguous model for spheres in 3-space is given by (Θ, χ), where $\Theta = \{\theta \in \mathbb{R}^4 \mid \tau(\theta)\}$ with $\tau(c, r) \equiv r > 0$ for $c = (c_x, c_y, c_z)$, and the characteristic function is given by

$$\chi(x, y, z, c, r) \equiv \|(x, y, z) - c\|^2 = r^2.$$

That is, c is the center and r is the radius of the sphere. In analogy to our discussion of spheres in 2-space we consider $\Theta' = \{\theta \in \mathbb{R}^9 \mid \tau'(\theta')\}$ where for two points $p_1 = (p_{1x}, p_{1y}, p_{1z})$, $p_2 = (p_{2x}, p_{2y}, p_{2z})$ and one normal $n = (n_x, n_y, n_z)$ we have

$$\tau'(p_1, n, p_2) \equiv \|n\|^2 = 1 \wedge p_1 \neq p_2,$$

and the intended semantics

$$\psi(c, r, p_1, n, p_2) \equiv \|p_1 - c\|^2 = r^2 \wedge \|p_2 - c\|^2 = r^2 \wedge (p_1 - c) \times n = 0.$$

We automatically generate Φ according to Theorem 4, drop the outermost block

$$\forall p_{1x} \forall p_{1y} \forall p_{1z} \forall p_{2x} \forall p_{2y} \forall p_{2z} \forall n_x \forall n_y \forall n_z$$

of quantifiers referrring to the variables of the intended model Θ', and apply generic quantifier elimination. We obtain within 30 ms the quantifier-free equivalent "true" subject to the conditions $A = \{n_x \neq 0, n_y \neq 0, p_{1x} \neq p_{2x}\}$.

Regular quantifier elimination on Φ does not succeed within reasonable time and space. Interestingly, generic quantifier elimination on some suitable $\tilde{\Phi}$ generated according to Theorem 7 delivers a slightly more complicated result here; compare the remarks after Example 11 about this. □

6   Tori

Technical applications frequently require the computation of a *parallel surface* P' to a given surface P in $\mathbb{R}^3$ [15, 8], i.e., for each point p' of P' the Euclidean distance $\inf_{p \in P} \|p' - p\|$ is constant. In [15] there is a parallel surface to a given aspheric surface computed in order to model the shrinkage of an aspheric lens. Reference [8] is concerned with another industrial application involving the automated estimation of the middle point position of a spheric milling cutter head. Here admissible head middle point positions lie on a parallel surface to the manufactured surface. In both these applications the runtime complexity of the basic operations on parallel surfaces is critical. One example for such basic operations is the minimal distance computation of a point to the given parallel surface.

In order to achieve acceptable runtime complexities the mathematical representations of the considered surfaces should be *simple*, which we want to make precise now. One established solution is to approximate the given surface by a spline of appropriate surface patches. One prominent example for such appropriate patches are toric patches [10]. Toric patches are simple at the first place due to the following invariance property: The parallel surface to a toric patch is a union of toric patches. Another advantage is the existence of simple formulas for the distance computation between a point and a toric patch. Last but not least a toric patch can be implicitly represented by a variety of degree four. This keeps small the number of parameters per patch needed to be stored in software systems. In a recent publication Schöne has shown that an arbitrary continuous function $f : D \to \mathbb{R}$ defined on a compact square $D \subseteq \mathbb{R}^2$ can be approximated by a spline constructed from generalized toric patches up to an arbitrary positive precision [14]. Generalized toric patches contain also degenerate tori, like cylinders or planes.

The work in [14] suffers, however, from an inherent drawback of approximating as a first step the surface by step functions, which introduces discontinuities and as a consequence does not lead to a sufficiently smooth approximation at the end. The need for better smooth approximations of surfaces with toric patches leads to the question under which conditions two toric patches have a $\mathbb{GC}^1$-*joint* [13]. A $\mathbb{GC}^1$-joint is a property of two smooth surfaces to have equal normals on their intersection curve. In that course, Martin introduces the notion of *principal patches* [10]. These are surfaces that are bounded by their lines of curvature. Martin shows that continuous joints between principal patches are possible along these curves. As an example for such surfaces he considers *cyclides*. These are surfaces where the lines of curvature are circles. The set of cyclides comprises as a special case the set of toric patches, where the lines of curvature are the meridians and the circles of latitude [16, 12]. Srinivas et al. construct surface approximations by smooth surfaces from cyclide patches with $\mathbb{GC}^1$-joints along the lines of curvature [16]. Mäurer and Krasauskas prove that a $\mathbb{GC}^1$-joint of two cyclide patches is possible at other lines than lines of curvature [11]. Schöne treats in [14] the following property of toric patches:

Conjecture 14. Between any two non-trivial toric patches obtained from distinct tori a $\mathbb{GC}^1$-joint with a regular intersection curve is possible only along their lines of curvature.

In other words, any regular intersection curve between the two patches is a line of curvature for each of them. Since the lines of curvature of tori are only meridians and circles of latitude, this implies that there is not much freedom to smoothly approximate surfaces exclusively using toric patches.

Conjecture 14 can be reduced to the following Conjecture, which is for fixed k tractable by automated deduction on the basis of the framework introduced in this paper:

Conjecture 15. There exists $k \in \mathbb{N}$ such that given a torus T, any family of pairwise distinct points $p_1, \ldots, p_k$ with normals $n_1, \ldots, n_k$ on T which do *not* lie on a common line of curvature uniquely determines T. □

Notice that finitely many points and normals on a common line of curvature, i.e. meridian or circle of latitude, cannot uniquely describe a torus: Given such points and normals one can freely choose the inner or outer radius, respectively.

Proof (of Conjecture 14 using Conjecture 15). Consider toric patches T_1 and T_2 obtained from different tori. Assume for a contradiction that the regular $\mathbb{GC}^1$ intersection curve L between T_1 and T_2 is not a line of curvature. We choose pairwise distinct $p_1, \ldots, p_k$ on L with normals $n_1, \ldots, n_k$. Since L is a $\mathbb{GC}^1$-joint curve the normals n_i apply to both T_1 and T_2. According to Lemma 15 these chosen points and normals describe a unique torus T. It follows that $T_1 = T$ and $T_2 = T$, a contradiction. □

Schöne proves a weaker variant of Conjecture 15 for $k = 4$ and $k = 5$ adding further requirements on the points $p_1, \ldots, p_k$ and normals $n_1, \ldots, n_k$, which then imply the uniqueness of T. The status of Conjecture 15 in its original form remains unclear in [14]. If Conjecture 15 is true, then there exists also a smallest $k \in \mathbb{N}$, which satisfies Conjecture 15. The original motivation of the present paper was the following conjecture:

Conjecture 16. The number $k = 4$ is the smallest possible choice in Conjecture 15. □

It is not hard to see that $k \leq 2$ cannot be a valid choice in Conjecture 15 [14]. Thus if Conjecture 15 is true for $k = 4$, then $k = 3$ is the only possible smaller choice, and in order to prove Conjecture 16 it suffices to give a counterexample for $k = 3$.

We are going to continue this discussion in Sect. 6.2, where we demonstrate how to apply our framework to prove Conjecture 15 and Conjecture 16. Before, we discuss in Sect. 6.1 some computations for circle rings, which are the 2-dimensional counterpart of tori and which we may expect to be computationally considerably easier.

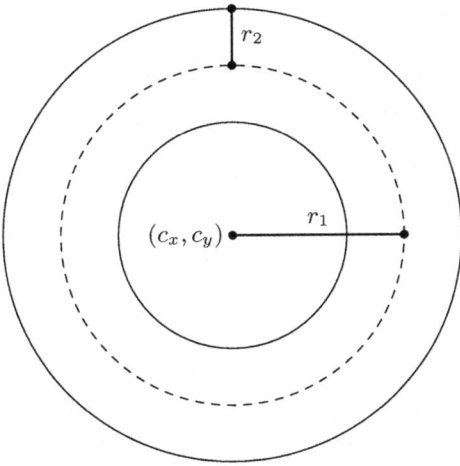

Fig. 3. Our reference model of a circle ring

6.1 Circle Rings in 2-Space

The family of circle rings in 2-space resembles the family of tori in 3-space: If
we intersect a torus with a plane, which has the rotation axis of the torus as
its normal, then we obtain as an intersection curve either the empty set, or one
circle or a *circle ring* consisting of two circles with a common center but distinct
radii. Second, the parallel curve to a circle ring patch is a union of circle ring
patches.

Consider $\Theta = \{\theta \in \mathbb{R}^4 \mid \tau(\theta)\}$ with $\tau(c_x, c_y, r_1, r_2) \equiv r_1 > r_2 \wedge r_2 > 0$.
A natural unambiguous model for a circle ring is given by interpreting $c = (c_x, c_y)$
as the common center and r_1 and r_2 as the radii of the two circles. We are,
however, going to use the following characteristic function instead:

$$\chi(x, y, c, r_1, r_2) \equiv \|(x, y) - c\|^2 = (r_1 - r_2)^2 \vee \|(x, y) - c\|^2 = (r_1 + r_2)^2.$$

Fig. 3 pictures the role of the two radii r_1 and r_2 in χ. This model exactly
corresponds to one common description of a torus, which we are going to use in
the next subsection.

We are now going to study how many points and normals we need to uniquely
describe a circle ring. Similar to our modeling of spheres in Examples 2 and 11
the following formula expresses for a circle ring with center $c = (c_x, c_y)$ and radii
r_1, r_2 that the point $p_i = (p_{ix}, p_{iy})$ lies on this circle ring and has the normal
$n_i = (n_{ix}, n_{iy})$ in this point:

$$\omega_i(c, r_1, r_2, p_i, n_i) \equiv \langle (-n_{iy}, n_{ix}) \mid p_i - c \rangle = 0 \wedge \|p_i + r_2 n_i - c\|^2 = r_1^2.$$

Notice that we assume n_i to be normalized and that p_i lies on the outer circle if
and only if n_i points into the direction of c.

We study the following statement: *Two points p_1, p_2 on a circle ring with two corresponding normals n_1, n_2 uniquely define the circle ring.*

For the simplest nontrivial case of two points with two normals we choose $\Theta_2' = \{\, \theta \in \mathbb{R}^8 \mid \tau_2'(\theta) \,\}$ with

$$\tau_2'(p_1, n_1, p_2, n_2) \equiv p_1 \neq p_2 \wedge \|n_1\|^2 = 1 \wedge \|n_2\|^2 = 1,$$

where obviously $p_1 \neq p_2 \equiv p_{1x} \neq p_{2x} \vee p_{1y} \neq p_{2y}$. Our intended semantics is given by

$$\psi_2(c, r_1, r_2, p_1, n_1, p_2, n_2) \equiv \omega_1 \wedge \omega_2.$$

With respect to the group of similarities we can choose the reference model $\Theta_0 = \{\, (0, 0, 1, r_2) \mid 0 < r_2 < 1 \,\}$, which represents circle rings in the origin with central radius 1. This yields $\sigma = [0/c_x, 0/c_y, 1/r_1]$. We automatically generate $\tilde{\Phi}$ according to Theorem 7 and drop the outermost universal quantifiers referring to p_1, n_1, p_2, and n_2.

Applying generic quantifier elimination yields after 15.5 s a quantifier-free equivalent containing 124 atomic formulas plus the assumptions

$$A = \{n_{1x}^2 + n_{1y}^2 - n_{2x}^2 - n_{2y}^2 \neq 0, n_{1x}n_{2y} - n_{1y}n_{2x} \neq 0, n_{1x} \neq 0\}.$$

The first assumption actually says $\|n_1\|^2 \neq \|n_2\|^2$. That is, the assumption quite trivially turns $\tilde{\Phi}$ true by contradicting the premise τ_2', which includes $\|n_1\|^2 = \|n_2\|^2 = 1$ and occurs as the antecedens of an implication. There are two possible approaches to avoid this: First, one can try switching off `rlqegenct` as discussed with Example 12. Second, one can universally quantify n_1 and n_2 in order to admit assumptions only on p_1 and p_2. Both these approaches lead to excessive computation times without satisfactory results.

We thus optimize our model using the following observation: Since circle rings in our reference model are invariant with respect to rotation around the origin we may without loss of generality assume that p_1 is located on the x-axis, say $p_1 = (1 + r_2, 0)$ and consequently $n_1 = (-1, 0)$. This additionally makes the assumption that p_1 is located on the outer circle, which will turn out not to be a restriction later. On the computational side this can be encoded by using another substitution $\rho = [1 + r_2/p_{1x}, 0/p_{1y}, -1/n_{1x}, 0/n_{1y}]$ and applying this to the matrix of $\tilde{\Phi}$ yielding $\tilde{\Phi}_\rho$. Generic quantifier elimination applied to $\forall n_2 \tilde{\Phi}_\rho$ followed by some automatic simplification using Gröbner basis methods [5] yields after only 120 ms altogether the following quantifier-free equivalent subject to the assumptions $A = \{p_{2x} \neq 0, p_{2y} \neq 0\}$:

$$p_{2x}^2 + p_{2y}^2 - 1 \leq 0 \vee p_{2x}^2 + p_{2y}^2 - 4 \geq 0 \vee (p_{2x}^3 + p_{2x}p_{2y}^2 \geq 0 \wedge p_{2x} < 0).$$

Since $p_{2x}^3 + p_{2x}p_{2y}^2 = p_{2x} \cdot \|p_2\|^2$ and $\|p_2\|^2 > 0$ due to A, the corresponding conjunction above simplifies to $p_{2x} \geq 0 \wedge p_{2x} < 0$, i.e., "false." The entire quantifier-free result can thus be rewritten as

$$\|p_2\| \leq 1 \vee \|p_2\| \geq 2.$$

Since in our model the center of the circle ring is the origin and the radius of the outer circle is $1 + r_2$ for $r_2 < 1$, the second condition states that p_2 does not lie on the circle ring at all thus trivially satisfying $\tilde{\Phi}_\rho$ by violating $\tau'\rho \wedge \psi_2\sigma\rho$, which is part of the antecedens of an implication in $\tilde{\Phi}_\rho$. From a less syntactic point of view, such points p_2 may occur as sub-vectors of $\theta' \in \Theta'_2$ but these θ' are not assigned to any parameter $\theta \in \Theta$ by the intended semantics. Similarly, for the first condition only the case $\|p_1\| = 0$ is relevant, which states that p_2 lies on the inner circle. Recall that we had positioned p_1 on the outer circle. Hence, p_1 and p_2 must lie on different circles, and we have learned a *forteriori* that our decision in ρ to place p_1 on the outer circle was not a loss of generality.

We are now going to discuss the two cases $p_{2x} = 0$ and $p_{2y} = 0$, which have been excluded in the course of the automatic introduction of A. This can be done by independently considering $\forall n_2 \tilde{\Phi}_\rho[0/p_{2x}]$ and $\forall n_2 \tilde{\Phi}_\rho[0/p_{2y}]$, respectively, and once more applying generic quantifier elimination to the considerably simpler problems obtained this way. Using the same simplification procedures as above, we obtain for the first case in less than 10 ms the following quantifier-free formula subject to the assumption $A = \{p_{2y} \neq 0\}$:

$$p_{2y} + 2 \leq 0 \vee p_{2y} - 2 \geq 0 \vee (p_{2y} + 1 \geq 0 \wedge p_{2y} - 1 \leq 0).$$

The first two constraints trivially satisfy the input formula by positioning p_2 beyond outer circle, and the conjunction yields our well-known condition that p_2 must lie on the inner circle. For the second case we obtain the assumption $A = \{p_{2x} \neq 0\}$ and the quantifier-free formula

$$p_{2x} + 2 \leq 0 \vee p_{2x} + 1 = 0 \vee p_{2x} - 1 \geq 0.$$

The first two constraints imply that p_2 does not lie on the circle ring at all. The last constraint implies $p_1 = (1 + r_2, 0) = p_2$, which is excluded by $\tau'_2\rho$. Hence, we have learned something new: Our chosen p_1 together with p_2 located on the x-axis cannot uniquely describe a circle ring. It is quite clear that the corresponding relevant condition in the general model, i.e., before the application of σ and ρ is that p_1, p_2 and the center c are not collinear. In both discussed special cases, the introduced assumptions A are not critical since they only exclude that p_2 is the origin, which would once more trivially satisfy $\forall n_2 \tilde{\Phi}_\rho$.

Notice that our obtained conditions greatly fit together with our introduction on the situation with tori: When obtaining a circle ring via intersection of a torus with a plane perpendicular to the rotation axis, the inner circle, the outer circle, and intersections of the circle ring with lines through the center are images of lines of curvature of that torus under this operation.

We conclude our discussion of circle rings with an interesting computational experiment: Most of the conditions obtained during our discussion above involved the distance between p_2 and the center of the circle ring. It is obvious that such conditions can be more naturally described when admitting to refer to the radius r_2 of the outer circle. We thus drop from $\forall n_2 \tilde{\Phi}_\rho$ the quantifier $\forall r_2$ and then apply generic quantifier elimination: This yields within 80 ms "true" subject to the following assumptions:

$$A = \{p_{2x}^2 + p_{2y}^2 - r_2^2 - 2r_2 - 1 \neq 0, p_{2x} \neq 0, p_{2y} \neq 0\}.$$

The first assumption $\|p_2\|^2 \neq (1 + r_2)^2$ states that p_2 must not lie on the outer circle, where we have put p_1. The third assumption states that p_2 does not lie on the x-axis, which reflects the necessary assumption on the original model that p_1 and p_2 must not be collinear with the center of the circle ring. Of course, this way we only know that these assumptions are sufficient while above we have automatically proved that they are both necessary and sufficient. It is, however, noteworthy that this style of automatic proving with its obvious weakness resembles techniques generally accepted for automatic proving of geometric theorems as, e.g., in the famous monograph by Chou [1].

6.2   Tori in 3-Space

The family of *tori* in $\mathbb{R}^3$ is given by $\Theta = \{\theta \in \mathbb{R}^8 \mid \tau(\theta)\}$ with

$$\tau(c, r, r_1, r_2) \equiv r_1 > r_2 \wedge r_2 > 0 \wedge \|r\| = 1 \wedge$$
$$(r_x > 0 \vee (r_x = 0 \wedge r_y > 0) \vee (r_x = 0 \wedge r_y = 0 \wedge r_z = 1)).$$

Here $c = (c_x, c_y, c_z)$ is the *center point* and $r = (r_x, r_y, r_z)$ is the direction of the *rotation axis* of the torus. We call r_1, $r_2 \in \mathbb{R}^+$ the *inner* and the *outer radius*, respectively. The *middle plane* of the torus is the plane with normal r containing c. The *middle circle* is the circle on the middle plane given by all points with distance r_1 to the center point.

According to [14] a point $p = (x, y, z) \in \mathbb{R}^3$ lies on the torus defined by $\theta = (c, r, r_1, r_2)$ if and only if using the notational convention $r' = (r'_x, r'_y, r'_z)$ we have

$$\chi(p, c, r, r_1, r_2) \equiv \exists r'(\|r'\|^2 = 1 \wedge \langle r \mid p + r_2 r' - c \rangle = 0 \wedge$$
$$\langle r \times r' \mid c - p \rangle = 0 \wedge \|p + r_2 r' - c\|^2 = r_1^2).$$

The model (Θ, χ) has 7 degrees of freedom, and it is unambiguous.

We now turn to first-order descriptions of normals on a torus in a given point. Denote by $S_3 \subseteq \mathbb{R}^3$ the unit sphere. A vector $n_i = (n_{ix}, n_{iy}, n_{iz}) \in S_3$ is a normal in the point $p_i = (p_{ix}, p_{iy}, p_{iz}) \in \mathbb{R}^3$ on a torus given by $\theta = (c, n, r_1, r_2)$ if and only if

$$\omega_i(c, r, r_1, r_2, p_i, n_i) \equiv \langle r \mid p_i + r_2 n_i - c \rangle = 0 \wedge$$
$$\langle r \times n_i \mid c - p_i \rangle = 0 \wedge \|p_i + r_2 n_i - c\|^2 = r_1^2.$$

For a number k of points $p_1, \ldots, p_k$ and normals $n_1, \ldots, n_k$ in these points we consider $\Theta'_k = \{\theta' \in \mathbb{R}^{6k} \mid \tau'_k(\theta')\}$ with $\tau'_k(p_1, n_1, \ldots, p_k, n_k)$ at least specifying

$$\bigwedge_{i=1}^{k} \|n_i\|^2 = 1.$$

An intended semantics for a torus model is then given by

$$\psi_k(c, r, r_1, r_2, p_1, n_1, \ldots, p_k, n_k) \equiv \bigwedge_{i=1}^{k} \omega_i.$$

With respect to the similarity group of $\mathbb{R}^3$ the reference family of tori can be represented by tori in the origin with the middle plane having the y-axis as normal. Additionally, in analogy to our model of circle rings, we can set $r_1 = 1$. This yields $\sigma = [0/c_x, 0/c_y, 0/c_z, 0/r_x, 1/r_y, 0/r_z, 1/r_1]$.

We now turn to a first-order formulation of the statement "points with normals lie on a common line of curvature of a torus." Recall from our discussion of circle rings in the previous section that we had automatically contained a corresponding condition by leaving certain variables unquantified.

In order to strictly apply our framework, however, suitable conditions have to be added to τ' in advance, and they have to be formulated in such a way that they refer exclusively to the coordinates of the points and normals under consideration but not to any torus. This is the content of the following lemma.

Lemma 17. *The points p_1, ..., p_k with corresponding normals n_1, ..., n_k lie on a line of curvature of some (not unique) torus T if and only if there exist $0 < \lambda \in \mathbb{R}$, a plane E, and $p \in E$ such that for each $1 \leq i \leq k$ we have*

(i) $p_i \in E$,
(ii) normal n_i is orthogonal to the normal of E, and
(iii) $p = \lambda n_i + p_i$.

Proof. For the direction from the left to the right it suffices to recall that the lines of curvature of tori are meridians and circles of latitude, both of which are circles. It is easy to see that (i)–(iii) are satisfied by any circle in 3-space and normals that lie on the same plane as the circle.

To prove the direction from the right to the left it is sufficient to construct a torus, for which the points p_1, ..., p_k and normals n_1, ..., n_k lie on a meridian. We choose the inner radius $r_1 = \lambda$ and the outer radius $r_2 = 2\lambda$. The center point of the torus c is then a (freely-chosen) point in E that has distance r_2 from p. The rotation axis is then $r = (p - c) \times n$, where n is the normal of E. It is not hard to see that the points p_1, ..., p_k and normals n_1, ..., n_k lie on a meridian for the specified torus. $\square$

For fixed k, the conditions in Lemma 17 can be written as a first-order formula over the reals such that we obtain the following τ_k' specifying Θ_k':

$$\tau_k'(p_1, n_1, \ldots, p_k, n_k) \equiv$$
$$\exists \lambda \exists p \exists n \exists d \bigwedge_{i=1}^{k} \left(\|n_i\| = 1 \wedge \langle p_i \mid n \rangle - d = 0 \wedge \langle n_i \mid n \rangle = 0 \wedge p = \lambda n_i + p_i \right).$$

From the existence of a quantifier elimination procedure for the reals it follows that these conditions can even be equivalently expressed without quantifiers.

On the basis of these preparations we can in theory automatically treat within our framework the following problems:

1. *Automatically prove Conjecture 16.* That is, show that the intended semantics ψ_3 of Θ_3' as specified by τ_3' is *not* adequate for (Θ, χ).

2. *Try to automatically prove Conjecture 15.* That is, check for increasing fixed
   $k = 4, 5, \ldots$ whether or not the intended semantics ψ_k of Θ_k' as specified by
   τ_k' is adequate for (Θ, χ).

Assume now that we succeed in proving Conjecture 16, i.e, the quantifier elimination returns "false" on the corresponding $\tilde{\Phi}$:

3. *Automatically construct a counterexample for Conjecture 16.* Here extended
   quantifier elimination would yield as a counterexample in addition to "false"
   three points with normals that uniquely describe a torus.

4. *Find additional conditions on three points and normals that do not lie on
   a line of curvature such that they uniquely describe a torus.* When adding
   first-order formulations of suitable conditions to τ_3' yielding τ_3'', we would
   have that ψ_3 becomes an adequate semantics of the corresponding Θ_3'' with
   respect to (Θ, χ).

Leaving certain variables unquantified, such additional condition can be obtained
either as quantifier-free equivalents or as assumptions with generic quantifier
elimination; compare our computations with circle rings. Technically, one can
possibly drop from τ_3' the conditions according to Lemma 17 and expect to
obtain these as well.

So far, we could not automatically solve any of the above problems in practice.
Detailed analyses of our test computations have shown that the main problem
is that the condition $\langle r \times n_i \mid c - p_i \rangle = 0$ in ω_i introduces a condition of
total degree 3, which leads to massive trial factorizations of polynomials during
quantifier elimination. An extension of the implementation of virtual substitution
in REDLOG to the cubic case [21] might considerably improve the situation.

Generally, we consider these open questions an interesting and practically
relevant challenge problem for automated deduction in geometry.

7   Conclusions

We have proposed a general framework to automatically prove uniqueness of geometric objects described by parameterizations intended by the user: On the basis
of an intended semantics, which relates the intended parameterization to a reference parameterization, we automatically generate first-order sentences over the
reals with ordering, which are true if and only if the intended parameterization
is adequate. This sentence can then be decided by real quantifier elimination or,
heuristically, by generic real quantifier elimination, where we possibly automatically derive non-degeneracy conditions missing the formulation. In a subsequent
quantifier elimination step we can in the positive case compute a quantifier-free
description of the characteristic function for our new parameterization. In the
negative case we can compute sample values witnessing the non-adequacy.

We have improved our basic framework by rigorously restricting to geometric objects in general position. It is noteworthy that our restrictions to general positions are systematically integrated into our framework in contrast to making ad hoc simplifications where the amount of human intelligence entering the automated deduction could hardly be specified or controlled.

The worst-case complexity of our approach is bounded by single exponential time.

We have given various computation examples with REDLOG in 2-space and 3-space using besides the present quantifier elimination services our procedures Φ and $\tilde{\Phi}$ implemented there for generating the input formulas.

The original motivation for our paper was a conjecture that toric patches can be joined smoothly only along lines of curvature, i.e. meridians or circles of latitude. We have reduced this conjecture to another one essentially tractable by automated deduction within our proposed framework: A torus is uniquely defined by finitely many points and corresponding normals if and only if these points do not lie on a common line of curvature. We have not yet succeeded in the automatic proof of this reduced conjecture but fully automatically treated the corresponding questions for circle rings in 2-space. Since we have exhibited that a proof or disproof of our conjecture has interesting practical consequences for spline approximation of parallel surfaces, we think that this is a challenging task for researchers in quantifier elimination and more generally for the entire *automated deduction in geometry* community.

Acknowledgments. The authors are grateful to R. Schöne and K. Donner for pointing at the existing research around toric spline interpolation and open problems in this area, as well as for useful references and many helpful discussions.

References

1. Chou, S.-C.: Mechanical Geometry Theorem Proving. Mathematics and its applications. D. Reidel Publishing Company, Dordrecht (1988)
2. Davenport, J.H., Heintz, J.: Real quantifier elimination is doubly exponential. Journal of Symbolic Computation 5(1-2), 29–35 (1988)
3. Dolzmann, A., Seidl, A., Sturm, T.: Efficient projection orders for CAD. In: Gutierrez, J. (ed.) Proceedings of the 2004 International Symposium on Symbolic and Algebraic Computation (ISSAC 2004), pp. 111–118. ACM Press, New York (2004)
4. Dolzmann, A., Sturm, T.: Redlog: Computer algebra meets computer logic. ACM SIGSAM Bulletin 31(2), 2–9 (1997)
5. Dolzmann, A., Sturm, T.: Simplification of quantifier-free formulae over ordered fields. Journal of Symbolic Computation 24(2), 209–231 (1997)
6. Dolzmann, A., Sturm, T., Weispfenning, V.: A new approach for automatic theorem proving in real geometry. Journal of Automated Reasoning 21(3), 357–380 (1998)
7. Dolzmann, A., Sturm, T., Weispfenning, V.: Real quantifier elimination in practice. In: Matzat, B.H., Greuel, G.-M., Hiss, G. (eds.) Algorithmic Algebra and Number Theory, pp. 221–247. Springer, Berlin (1998)

8. Jüttler, B., Sampoli, M.L.: Hermite interpolation by piecewise polynomial surfaces with rational offsets. Computer-Aided Geometric Design 17, 361–385 (2000)

9. Loos, R., Weispfenning, V.: Applying linear quantifier elimination. The Computer Journal 36(5), 450–462 (1993)

10. Martin, R.R.: Principal patches. a new class of surface patch based on differential geometry. In: Eurographics 1983, pp. 47–55. North Holland, Amsterdam (1984)

11. Mäurer, C., Krasauskas, R.: Joining cyclide patches along quartic boundary curves. In: Dæhlen, M., Lyche, T., Schumaker, L.L. (eds.) Proceedings of the International Conference on Mathematical Methods for Curves and Surfaces II, Lillehammer, pp. 359–366. Vanderbilt University, Nashville (1998)

12. Pratt, M.J.: Cyclides in computer-aided geometric design. Computer-Aided Geometric Design 7, 221–242 (1990)

13. Prautzsch, H., Boehm, W., Paluszny, M.: Bézier and B-Spline Techniques. Springer, Berlin (2002)

14. Schöne, R.: Torische Splines. Doctoral dissertation, Department of Computer Science and Mathematics. University of Passau, Germany, D-94030 Passau, Germany (2007)

15. Schöne, R., Hintermann, D., Hanning, T.: Approximation of shrinked aspheres. In: Gregory, G.G., Howard, J.M., Koshel, R.J. (eds.) International Optical Design Conference 2006 (Proceedings of SPIE-OSA). Proceedings of SPIE, vol. 6342. SPIE, Bellingham (2006)

16. Srinivas, Y.L., Kumar, V., Dutta, D.: Surface design using cyclide patches. Computer-Aided Design 28, 263–276 (1996)

17. Sturm, T.: Real Quantifier Elimination in Geometry. Doctoral dissertation, Department of Mathematics and Computer Science. University of Passau, Germany, D-94030 Passau, Germany (December 1999)

18. Sturm, T.: Reasoning over networks by symbolic methods. Applicable Algebra in Engineering, Communication and Computing 10(1), 79–96 (1999)

19. Sturm, T., Weispfenning, V.: Computational geometry problems in Redlog. In: Wang, D. (ed.) ADG 1996. LNCS (LNAI), vol. 1360, pp. 58–86. Springer, Heidelberg (1998)

20. Weispfenning, V.: The complexity of linear problems in fields. Journal of Symbolic Computation 5(1-2), 3–27 (1988)

21. Weispfenning, V.: Quantifier elimination for real algebra—the cubic case. In: Proceedings of the International Symposium on Symbolic and Algebraic Computation, Oxford, England (ISSAC 1994), pp. 258–263. ACM Press, New York (1994)

22. Weispfenning, V.: Quantifier elimination for real algebra—the quadratic case and beyond. Applicable Algebra in Engineering Communication and Computing 8(2), 85–101 (1997)

23. Weispfenning, V.: Simulation and optimization by quantifier elimination. Journal of Symbolic Computation 24(2), 189–208 (1997)

Formalizing Projective Plane Geometry in Coq*

Nicolas Magaud, Julien Narboux, and Pascal Schreck

LSIIT UMR 7005 CNRS - Université de Strasbourg

Abstract. We investigate how projective plane geometry can be formalized in a proof assistant such as Coq. Such a formalization increases the reliability of textbook proofs whose details and particular cases are often overlooked and left to the reader as exercises. Projective plane geometry is described through two different axiom systems which are formally proved equivalent. Usual properties such as decidability of equality of points (and lines) are then proved in a constructive way. The duality principle as well as formal models of projective plane geometry are then studied and implemented in Coq. Finally, we formally prove in Coq that Desargues' property is independent of the axioms of projective plane geometry.

Keywords: formalization, projective geometry, duality, Coq.

1 Introduction

This paper deals with formalizing projective plane geometry. Projective plane geometry can be described by a fairly simple set of axioms. However it captures the main aspects of plane geometry, especially perspective. It is a good candidate to be formalized in a proof assistant. Most of the description and proofs are available in textbooks such as [9,3]. However, in most books, many lemmas are considered trivial and many proofs are left to the reader. Building a formal development in a proof assistant allows for more flexibility. If required, axioms can be changed easily and proofs can be rechecked automatically by the system. Such changes may only require minor rewriting of the proofs by the user. In all cases, the proofs are computer-verified, which dramatically increases their reliability compared to paper-and-pencil proofs.

This formalization is not only interesting in itself. It also allows to evaluate the adequacy of a proof assistant such as Coq to develop a formal theory and to build some models of this theory. More significantly, we formalize projective plane geometry because we are interested in building reliable and robust constraint solving programs (see [16,15]). Indeed, in geometric constraint solving, handling the numerous particular cases is crucial to ensure robustness. Detecting whether a configuration is degenerated or not requires theorem proving [30]: which theorems are required and how to prove them is among the issues we want to address. As shown in [21], point-line incidences in the projective plane are sufficient to express usual geometric constraints.

---

* This work is partially supported by the ANR project Galapagos.

T. Sturm and C. Zengler (Eds.): ADG 2008, LNAI 6301, pp. 141–162, 2011.

Finally, as computer scientists, we are interested in the effectiveness of proofs in order to extract programs from these proofs. The Coq proof assistant [8,1] implements a constructive logic and allows program extraction from constructive proofs. Therefore, it is the perfect tool to carry out a constructive formalization.

In this paper, we formalize the theory of projective plane geometry and we build models of this theory. In a subsequent paper [18], we revisit and generalize the axiom system for projective geometry in a at least 3 dimensional setting using flats and ranks and prove Desargues' property holds in that case.

Related Work. Proof assistants have already been used in the context of geometry. The task consisting in mechanizing Hilbert's *Grundlagen der Geometrie* has been partially achieved. A first formalization using the Coq proof assistant was proposed by Christophe Dehlinger, Jean-François Dufourd and Pascal Schreck [11]. This first approach was realized in an intuitionist setting, and concluded that the decidability of point equality and collinearity is necessary to check Hilbert's proofs. Another formalization using the Isabelle/Isar proof assistant [26] was performed by Jacques Fleuriot and Laura Meikle [19]. Both formalizations have concluded that, even if Hilbert has done some pioneering work about formal systems, his proofs are in fact not fully formal, in particular degenerated cases are often implicit in the presentation of Hilbert. The proofs can be made more rigorous by machine assistance. Frédérique Guilhot realized a large Coq development about Euclidean geometry following a presentation suitable for use in french high-school [13]. In [24,25], Julien Narboux presented the formalization and implementation in the Coq proof assistant of the area decision procedure of Chou, Gao and Zhang [6] and a formalization of foundations of Euclidean geometry based on Tarski's axiom system [33,31]. In [12], Jean Duprat proposes the formalization in Coq of an axiom system for compass and ruler geometry. Regarding formal proofs of algorithms in the field of computational geometry, we can cite David Pichardie and Yves Bertot [27] for their formalization of convex hulls algorithms in Coq as well as Laura Meikle and Jacques Fleuriot [20] for theirs in a Hoare-like framework in Isabelle. Several papers introduce methods for automatic proof in projective geometry, e.g. [29,17]. Our work is different because we perform *interactive* proofs in projective geometry. Our approach is only slightly automated, but we can deal with the degenerated cases by careful study in the proof assistant whereas these cases are ignored in [29]. In addition we can deal with theorems which are not stated as a geometric construction, which is a limitation of approaches based on the Wu's method and the area method.

Notations. Most Coq notations, which are really close to mathematical ones, will be explained along the course of the paper. The negation is noted ~. The most awkward notation for the reader not accustomed to Coq is the curly-brackets notation for constructive existential quantification over the sort Type. For instance, the formula `forall l:Line, {X:Point | ~Incid X l}` expresses that $\forall l : \text{Line}, \exists X : \text{Point}, \neg X \in l$.

Outline. The paper is organized as follows. In Sect. 2, we present the axioms for projective plane geometry and their description in the Coq proof assistant. In Sect. 3, we study the duality between points and lines. Section 4 deals with finite and infinite models for projective plane geometry. In Sect. 5 we build both desarguesian and non-desarguesian models.

2   Axioms

2.1   A First Set of Axioms

We assume that we have two kinds of objects which we call points and lines. We also assume that we have a relation ($\in$) between elements of these two sets. Projective plane geometry can be described using the six axioms presented in Fig. 1.

The first two axioms deal with existence of points and lines. We choose not to require points (resp. lines) to be distinct in axiom 'Line Existence' (resp. 'Point Existence'). If the points (resp. lines) are equal, the line (resp. the point) still

Axiom Line Existence

$$\forall A \ B : \text{Point}, (\exists l : \text{Line}, A \in l \wedge B \in l)$$

Axiom Point Existence

$$\forall l \ m : \text{Line}, (\exists A : \text{Point}, A \in l \wedge A \in m)$$

Axiom Line Uniqueness

$$\forall A \ B : \text{Point}, A \neq B \Rightarrow \forall l \ m : \text{Line}, A \in l \wedge B \in l \wedge A \in m \wedge B \in m \Rightarrow l = m$$

Axiom Point Uniqueness

$$\forall l \ m : \text{Line}, l \neq m \Rightarrow \forall A \ B : \text{Point}, A \in l \wedge A \in m \wedge B \in l \wedge B \in m \Rightarrow A = B$$

Definition (distinct4)

$$\text{distinct4} \ A \ B \ C \ D \equiv A \neq B \ \wedge \ A \neq C \ \wedge \ A \neq D \ \wedge \ B \neq C \ \wedge \ B \neq D \wedge C \neq D$$

Axiom Four Points

$$\exists A : \text{Point}, \exists B : \text{Point}, \exists C : \text{Point}, \exists D : \text{Point},$$
$$\text{distinct4} \ A \ B \ C \ D \wedge$$
$$(\forall l : \text{Line}, (A \in l \wedge B \in l \Rightarrow C \notin l \wedge D \notin l) \wedge$$
$$(A \in l \wedge C \in l \Rightarrow B \notin l \wedge D \notin l) \wedge$$
$$(A \in l \wedge D \in l \Rightarrow B \notin l \wedge C \notin l) \wedge$$
$$(B \in l \wedge C \in l \Rightarrow A \notin l \wedge D \notin l) \wedge$$
$$(B \in l \wedge D \in l \Rightarrow A \notin l \wedge C \notin l) \wedge$$
$$(C \in l \wedge D \in l \Rightarrow A \notin l \wedge B \notin l))$$

Fig. 1. A first axiom system for projective plane geometry

exists: actually there potentially exists an infinity of lines (resp. points). This design choice follows a general rule in formal geometry: it is crucial to consider statements which are as general as possible.

The next two axioms deal with uniqueness of the above defined line and point. These axioms hold only if the two points (resp. lines) are distinct. As suggested in [2], axioms 'Point Uniqueness' and 'Line Uniqueness' can be merged into a more convenient axiom with no negation. This axiom is classically equivalent to the conjunction of the two others:

Axiom Uniqueness

$$\forall A\ B : \text{Point}, \forall l\ m : \text{Line},$$
$$A \in l \Rightarrow B \in l \Rightarrow A \in m \Rightarrow B \in m \Rightarrow A = B \vee l = m$$

Finally, axiom 'Four points' states that there exists at least four distinct points, no three of them being collinear. This means dimension is at least 2. Together with axiom 'Point Existence' which expresses that the dimension is at most 2 (two lines always intersect), it imposes that the dimension of this projective space is exactly 2. The formalization of this axiom system is straightforward, but from a practical point of view, proofs in most textbooks often use some variants of this system. To ease mechanization of proofs, we formalize the equivalence between these systems.

2.2   Another Axiom System for Projective Plane Geometry

Another Non-degeneracy Axiom. Axiom 'Four Points' states a non-degeneracy condition, namely that the projective space we consider is not reduced to a single line. This can be expressed in another way through two new axioms:

Axiom Three Points

$$\forall l : \text{Line}, \exists ABC : \text{Point}, A \neq B \wedge B \neq C \wedge A \neq C \wedge A \in l \wedge B \in l \wedge C \in l$$

Axiom Lower Dimension

$$\exists l_1 : \text{Line}, \exists l_2 : \text{Line}, l_1 \neq l_2$$

The first axiom expresses that each line contains at least three points; the second one states that there exist two distinct lines.

We prove that axiom 'Four points' can be replaced by axiom 'Three points' and axiom 'Lower dimension' in the system defined in the previous section and vice-versa. Both settings share the following axioms: Line Existence, Point Existence, Uniqueness. In mathematics textbooks, the equivalence of these two sets of axioms is usually presented as a remark.[1] For instance in [3], the proof is left to the reader. In a proof assistant such as Coq, these proofs have to be made explicit and proving them formally requires some technical work mostly related to handling the numerous configurations of points. The basic principles of the proof are presented in Appendix A.

---

[1] Proving this in Coq requires more than 1000 lines of proof.

2.3   Implementation in Coq

We formalize the previous definitions in the Coq proof assistant [1,8]. To do so, we take advantage of the modules and functors of Coq. Modules [7] allow to define parametrized theory and to put together types and definitions into a module structure. This enhances the re-usability of developments, by providing a formal interface for such a structure. In addition, functors can be used to connect module types to one another.

Modules and Projective Plane. Our first module `PreProjectivePlane` contains axioms dealing with point (resp. line) existence and uniqueness. From that

```
Module Type ProjectivePlane.

Parameter Point: Set.
Parameter Line: Set.
Parameter Incid : Point -> Line -> Prop.

Axiom incid_dec : forall (A:Point)(l:Line), {Incid A l} + {~Incid A l}.

(* Line Existence : any two points lie on a unique Line *)

Axiom a1_exist : forall (A B :Point),
                           {l:Line | Incid A l /\ Incid B l}.

(* Point Existence : any two lines meet in a unique point *)
Axiom a2_exist : forall (l1 l2:Line),
                           {A:Point | Incid A l1 /\ Incid A l2}.

Axiom uniqueness : forall A B :Point, forall l m : Line,
  Incid A l -> Incid B l  -> Incid A m -> Incid B m -> A=B \/ l=m.


(* Four points : there exist four points with no three collinear *)
Axiom a3: {A:Point & {B :Point & {C:Point & {D :Point |
  (forall l :Line, distinct4 A B C D/\
      (Incid A l /\ Incid B l -> ~Incid C l /\ ~Incid D l)
   /\ (Incid A l /\ Incid C l -> ~Incid B l /\ ~Incid D l)
   /\ (Incid A l /\ Incid D l -> ~Incid C l /\ ~Incid B l)
   /\ (Incid C l /\ Incid B l -> ~Incid A l /\ ~Incid D l)
   /\ (Incid D l /\ Incid B l -> ~Incid C l /\ ~Incid A l)
   /\ (Incid C l /\ Incid D l -> ~Incid B l /\ ~Incid A l))}}}}.

End ProjectivePlane.
```

Fig. 2. The module type with axioms required to describe a projective plane. The incidence relation ($\in$) is noted `Incid` in our Coq development. `{Incid A l}` + `{~Incid A l}` expresses that we know *constructively* that $A \in l \vee \neg A \in l$.

we derive some basic properties, including uniqueness of a line (resp. of a point), from the general uniqueness axiom. Then, on top of `PreProjectivePlane`, we build two modules `ProjectivePlane` which contains axiom 'Four points' and `ProjectivePlane'` which contains axiom 'Three points' and axiom 'Lower dimension'. A theory is of type `ProjectivePlane` if it contains all the notions presented in Fig.2. The two module types `ProjectivePlane` and `Projective-Plane'` are connected through two functors `Back` and `Forth` which prove the equivalence of these two axiom systems when the axioms 'Line Existence' and 'Point Existence' as well as Uniqueness are available. Figure 2 sums up the module type for projective plane geometry and Fig. 3 presents the global organization of the development.

Deciding Equality. From the assumption that incidence is decidable:

$$\forall A : \text{Point}, \forall l : \text{Line}, (A \in l \vee \neg A \in l),$$

we prove that point (resp. line) equality is decidable. The proofs of decidability for point (resp. line) equality are intuitionist, in the sense that they do not use the excluded middle property. Details of these proofs are available in Appendix B.

From these basic axioms, we can consider proving some theorems about projective plane geometry. For instance, we prove that if we consider lines as set of points, there always exists a bijection between two lines (see Appendix E for details). In order to improve genericity, we show that the well-known principle of duality between point and line can be derived in Coq. It allows us to prove automatically half of the theorems of interest from the proofs of their dual counterparts.

3   Duality

3.1   Principle of Duality

It is well known that projective geometry enjoys a principle of duality, namely that every definition remains significant and every theorem remains true, when we interchange the concepts *Point* and *Line*. But as we exchange points and lines, predicates must be exchanged with their dual as well. For example, the collinearity property, i.e. *col A B C* $\equiv \exists l : \text{Line}, A \in l \wedge B \in l \wedge C \in l$ must be replaced by the concurrency property i.e. *meet a b c* $\equiv \exists L : \text{Point}, a \ni L \wedge b \ni L \wedge c \ni L$. To formalize this principle, we make use of the module system of Coq [1,7]. In practice, we consider the module type `ProjectivePlane'` defined in the previous section and we build a functor from `ProjectivePlane'` to itself in which we map points to lines and lines to points:

```
Module swap (M: ProjectivePlane') : ProjectivePlane'.
Definition Point := M.Line.
Definition Line  := M.Point.
Definition Incid := fun (x:Point) (y:Line) => M.Incid y x.
...
```

To build this functor we need to show that the dual of each axiom holds. It is clear that the axioms of existence and uniqueness of lines are the dual of the axioms for existence and uniqueness of points:

```
Definition a1_exist  := M.a2_exist.
Definition a1_unique := M.a2_unique.
Definition a2_exist  := M.a1_exist.
Definition a2_unique := M.a1_unique.
```

To prove the dual version of axiom 'Three points' and axiom 'Lower dimension' it is necessary to use the other axioms. Appendix C provides the detailed proof of the fact that incidence geometry is a dual of itself and Fig. 3 a summary of the organization of the development.

3.2 Applications

Formalizing this principle of duality leads to an interesting theoretical result. In addition, this principle is also useful in practice. For every theorem we prove, we can easily derive its dual using our functor swap. For instance,

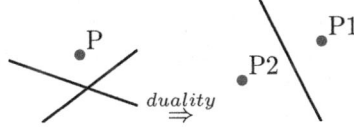

from the lemma outsider stating that for every couple of lines, there is a point which is not on these lines, we can derive its dual automatically: for every couple of points, there is a line not going through these points.

```
Module Example (M': ProjectivePlane').

Module Swaped := swap M'.
Export M'.

Module Back := back.back Swaped.
Module ProjectivePlaneFacts_m := decidability.decidability Back.

Lemma dual_example :
forall P1 P2 : Point,{l : Line | ~ Incid P1 l /\ ~ Incid P2 l}.
Proof.
apply ProjectivePlaneFacts_m.outsider.
Qed.

End Example.
```

So far, we focused on axiom systems and formal proofs. The next step is to check whether well-known models verify our axioms for projective plane geometry.

4 Models

In order to prove formally that our sets of axioms are consistent, we build some models. We build both finite and infinite models: among them the smallest projective plane and an infinite model based on homogeneous coordinates.

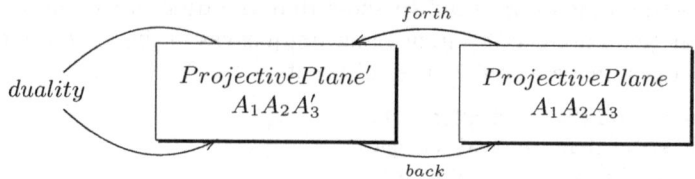

Fig. 3. A modular organization. Arrows represent functors and boxes represent modules types.

4.1 Finite Models

Following Coxeter's notation [9], a finite projective geometry is written $PG(a, b)$ where a is the number of dimensions, and given a point on a line, b is the number of other lines through the point. We build two finite models: $PG(2, 2)$ and $PG(2, 5)$. $PG(2, 2)$ is the smallest projective plane and is also known as Fano's plane.

Fano's Plane. In two dimensions, we can easily build the model with the least number of points and lines: 7 each. This model is called Fano's plane. On the figure, points are simply represented by points, whereas lines are represented by six segments and a circle (DEF).

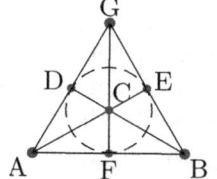

In order to formalize this model, we define a module `FanoPlane` of type `ProjectivePlane`. The typing system of Coq will ensure that our definitions are really instances of the abstract definition of a projective plane.

The set of points is defined by an inductive[2] type with 7 constructors and the set of lines as well:

```
Inductive ind_point : Set := A | B | C | D | E | F | G.
Inductive ind_line  : Set := ABF | BCD | CAE | ADG | BEG | CFG | DEF.

Definition point : Set := ind_point.
Definition line  : Set := ind_line.
```

The incidence relation is given explicitly by its graph:

```
Definition incid_bool : Point -> Line -> bool := fun P L =>
 match (P,L) with
  (A,ABF) | (A,CAE) | (A,ADG) | (B,BCD) | (B,BEG) | (B,ABF)
 |(C,BCD) | (C,CAE) | (C,CFG) | (D,BCD) | (D,ADG) | (D,DEF)
 |(E,CAE) | (E,BEG) | (E,DEF) | (F,ABF) | (F,DEF) | (F,CFG)
 |(G,ADG) | (G,CFG) | (G,BEG)  => true
 | _ => false
end.
```

---

[2] Note that this type is not really inductive, but sum types are defined in Coq using a special case of the general concept of inductive types.

The proofs of existence and uniqueness are performed by case analysis. Note that in order to prove the axioms of uniqueness, we must prove that for every couple of points (resp. lines) there is a unique line (resp. point). This creates $7^2 = 49$ cases. For each of these cases, we have to perform a case analysis on the lines, this produces again 49 cases, leading to a total of 2401 cases. The proof is computed easily by Coq.

PG(2,5). We follow [9] and build another model of the projective plane which is still finite but larger than Fano's plane. This model is called $PG(2,5)$. It contains 31 points and as many lines. The incidence relation is given on Table 1 in Appendix D. From the technical view of the formalization, this model is harder to build than Fano's plane because the proof produces 923 521 cases[3]. However, the proofs of these cases can be automated. The total size of the proof object generated by Coq (a term of the calculus of inductive constructions) is 7 Mo.

4.2   Infinite Model: Homogeneous Coordinates

To build an infinite model of projective geometry we use homogeneous coordinates introduced by August Ferdinand Möbius. We present our formalization in the context of the projective plane, but it can be easily generalized to any other dimension. The homogeneous coordinates of a point (resp. of a line) of a projective plane is a triple of numbers which are not all zero. These numbers are elements of any commutative field of characteristic zero. Two triples which are proportional are considered as equal: for any $\lambda \neq 0$, $(x_1, x_2, x_3) = (\lambda x_1, \lambda x_2, \lambda x_3)$.

To formalize this notion in Coq it would be natural to define pseudo-points as triple of elements of a field and then define points (resp. lines) as the equivalence classes of proportional non-zero triple in this field. Unfortunately, defining a type by quotient is something difficult to do in the calculus of inductive constructions used by Coq [5]. Therefore, we choose to define the quotient type directly by representing the classes of points and lines by a normal form. Points and lines are represented by their triple of coordinates such that the last non zero coordinate is 1. Consider a point (x_1, x_2, x_3). If $x_3 \neq 0$ we can represent it by $(x_1/x_3, x_2/x_3, 1)$. If $x_3 = 0$, we perform case distinction on x_2. If $x_2 \neq 0$ we can represent it by $(x_1/x_2, 1, 0)$, else we represent it by $(1, 0, 0)$. This definition can be formalized in Coq using the following inductive type where F is the type of the elements of our field and P0, P1 and P2 are the constructors for the three different cases:

---

[3] In [9], the proof given is the following: "we observe that any two residues are found together in just one column of the table (see on page 161), and that any two columns contain just one common number". This amounts to checking, for more than 400 different configurations, whether two sets of six elements have only one common element. In such a case, mechanized theorem proving is the best way to ensure correctness.

```
Inductive Point : Set :=
 | P2 : F -> F -> Point (* (x1,x2,1) *)
 | P1 : F -> Point      (* (x1,1 ,0) *)
 | P0 : Point.          (* (1 ,0 ,0) *)
```

The second and third constructors correspond to ideal points (points at infinity).

The incidence relation (noted Incid in Coq and $\in$ in this paper) can then be defined as the inner product of a point and line. The definition of the inner product can be made more generic by using triples, instead of giving a definition distinguishing each of the 3×3 cases.

To do this, we define two functions, one to transform a point into a triple of coordinates, and another one to normalize a triple of coordinates to obtain a point. We can then prove two lemmas which state that our definitions are consistent:

```
Lemma triple_point :
 forall P : Point, point_of_triple (triple_of_point P) = P.
```

```
Lemma point_triple :
 forall a b c : F, (a,b,c) <> (0,0,0) ->
 exists l, triple_of_point (point_of_triple (a,b,c)) = (a*l,b*l,c*l).
```

```
Lemma point_of_triple_functionnal:
 forall a b c l : F, (a,b,c) <> (0,0,0) -> l <> 0 ->
 point_of_triple(a,b,c) = point_of_triple(a*l,b*l,c*l).
```

The inner product and incidence relations can then be defined as:

```
Definition inner_product_triple A B :=
match (A,B) with
 ((a,b,c),(d,e,f)) => a*d+b*e+c*f
end.
```

```
Definition Incid : Point -> Line -> Prop := fun P L =>
 inner_product_triple (triple_of_point P) (triple_of_line L) = 0.
```

Now, we need to prove that the axioms of a projective plane hold in this setting. The proof of the decidability of Incid and of axioms (Three Points) and (Lower Dimension) are straightforward. For the uniqueness axiom, after unfolding of definitions, the problem reduces to a goal involving equations such as the following ones:

$$
\begin{array}{l}
r * r_5 + r_0 * r_6 + 1 = 0 \\
r * r_3 + r_0 * r_4 + 1 = 0 \\
r_1 * r_5 + r_2 * r_6 + 1 = 0 \\
r_1 * r_3 + r_2 * r_4 + 1 = 0
\end{array} \Rightarrow (r = r_1 \wedge r_0 = r_2) \vee (r_3 = r_5 \wedge r_4 = r_6).
$$

Using the following equivalences considered as rewrite rules, we can convert this goal into an ideal-membership problem which can be solved using the Gröbner bases tactic developed by Loïc Pottier [10]:

$$\forall ab, \qquad a = b \qquad \Leftrightarrow \qquad a - b = 0$$
$$\forall ab, (a = 0 \lor b = 0) \Leftrightarrow \qquad ab = 0$$
$$\forall ab, (a = 0 \land b = 0) \Leftrightarrow a^2 + b^2 = 0.$$

This tactic provides automation to solve algebraic goals which otherwise would be tedious to prove interactively. Proofs achieved by the Gröbner basis tactic require less than 2 seconds of computation except one which requires about a minute.

Finally, for the existence axioms, we need to define the line passing through two points (resp. the point at the intersection of two lines).

5   Desarguesian and Non-desarguesian Models

Desargues' property is among the most fundamental properties of projective geometry since in the projective space Desargues' property becomes a theorem and consequently all the projective spaces arise from a division field. In this section, we formalize two models showing on the one hand, that Desargues' property is compatible with the axioms of projective geometry and, on the other hand that it is independent of them. Let's first recall Desargues' statement in projective geometry.

5.1   Desargues' Property

Desargues' property states that: *Let E be a projective space and A,B,C,A',B',C' be points in E, if the three lines joining the corresponding vertices of triangles ABC and A'B'C' all meet in a point O, then the three intersections of pairs of corresponding sides α, β and γ lie on a line.*

If E is at least of dimension three, Desargues' property always holds. In [18], we have formalized this theorem in Coq. If E is of dimension two, Desargues'

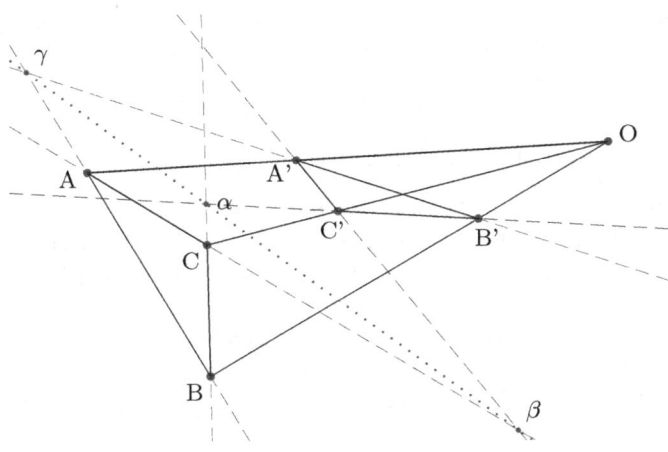

Fig. 4. Desargues' Property

property is independent from all the projective plane geometry axioms. We first show that it is not contradictory with these axioms by formally proving it holds in Fano's plane. Then, we build a model (Moulton's plane) in which all projective plane geometry axioms hold but Desargues' property does not hold. This is achieved by making explicit a configuration for which Desargues' property is not satisfied. Overall this shows the independence of Desargues's theorem from the axioms of projective plane geometry, which can be regarded as the starting point of non-desarguesian geometry [4].

5.2   Fano's Plane Is Desarguesian

At first sight, proving Desargues' property in Fano's plane seems to be straightforward to achieve by case analysis on the 7 points and 7 lines. However, this requires handling numerous cases[4] including many configurations which contradict the hypotheses.

To formalize the property we make use of two kinds of symmetries, a symmetry of the theory and a symmetry of the statement.

Symmetry of the Statement. We first study the special case where the point O of Desargues' configuration corresponds to A, and the line OA corresponds to ADG, OB to CAE and OC to ADF. Then as Desargues' statement is symmetric by permutation of the three lines which intersect in O, we can formalize a proof of slightly more general lemma `desargues_from_A` where the point O of Desargues' configuration still corresponds to A but the three lines intersecting in O are universally quantified.

Symmetry of the Theory. The theory of Fano's plane is invariant by permutation of points. It means that, even if it is not obvious from the figure in Sect. 4.1, all the points play the same role: if (A, B, C, D, E, F, G) is a Fano's plane then (B, C, D, F, E, G, A) is one as well. We formalize this by building a functor from Fano's theory to itself which permutes the points. Using this functor and `desargues_from_A`, we show that Desargues' property holds for any choice for O among the 7 points of the plane.

5.3   Independence of Desargues' Property

Moulton Plane and its Projective Counterpart. Moulton plane [23] is an affine plane in which lines with a negative slope are bent (i.e. the slope is doubled) when they cross the y-axis. It can be easily extended into a projective plane.

Moulton plane is an incidence structure which consists of a set of points P, a set of lines L, and an incidence relation between elements of P and elements of L. Points are denoted by couples $(x, y) \in \mathbb{R}^2$. Lines are denoted by

---

[4] The most naïve approach would consider 7^7 cases, even with careful analysis it remains untractable to prove all the cases without considering symmetries.

$(m, b) \in (\mathbb{R} \cup \infty) \times \mathbb{R}$ (where m is the slope - ∞ for vertical lines - and b the y-intercept). The incidence relation is defined as follows:

$$(x, y) \in (m, b) \iff \begin{cases} x = b & \text{if } m = \infty \\ y = mx + b & \text{if } m \geq 0 \\ y = mx + b & \text{if } m \leq 0, x \leq 0 \\ y = 2mx + b & \text{if } m \leq 0, x \geq 0. \end{cases}$$

This incidence structure verifies the properties of an affine plane. It can be turned into a projective plane through the following process.

- We extend P with points at infinity (one *direction point* for each possible slope, including the vertical one); therefore P is $(\mathbb{R} \times \mathbb{R}) \cup \mathbb{R}$.
- We extend the set L of affine lines with a new one which connects all points at the infinite; therefore L is $((\mathbb{R} \cup \infty) \times \mathbb{R}) \cup \infty$.
- We finally extend the incidence relation in order to have all *direction points* and only them incident to the infinite line. We also extend each affine line with a direction point (the one bearing its slope).

This construction leads to a projective plane. The whole process is formally described in Coq and we show that all the axioms of projective plane geometry presented in Sect. 2 hold. Most proofs on real numbers rely on using Gröbner basis computation in Coq as already used in Sect. 4.2.

A Configuration of Desargues where the Theorem Does Not Hold. We build a special configuration of Desargues for which the property does not hold. This can be achieved in a very algebraic point of view using only coordinates and equations for lines. We first present it that way and then show on a figure why Desargues' property does not hold for our configuration.

Let's consider 7 points: $O(-4, 12)$, $A(-8, 8)$, $B(-5, 8)$, $C(-4, 6)$, $A'(-14, 2)$, $B'(-7, 0)$ and $C'(-4, 3)$. We then build the points $\alpha(-3, 4)$, $\beta(6/11, 38/11)$ and $\gamma(-35, 8)$ which are respectively at the intersection of (BC) and $(B'C')$, (AC) and $(A'C')$ and (AB) and $(A'B')$. Then we can check using automated procedures on real numbers computation that there exists no line in Moulton plane which is incident to these 3 points α, β and γ.

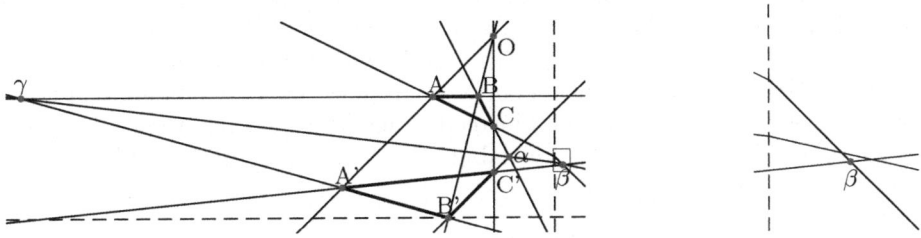

Fig. 5. Counter example to Desargues' theorem in Moulton's plane

Overall Desargues' property does not hold in this configuration because only some of the lines are bent. Especially, of the two lines used to build β, one of them is a straight line ($A'C'$) and the other one (AC) is bent. That is what prevents the three points α, β and γ from being on the same line.

Proofs studied in previous sections illustrate how combining *automated* and *interactive* theorem proving can be successful.

6   Conclusion and Future Work

In this paper, we have shown how projective plane geometry can be formalized in Coq using two different axiom systems. We proved them equivalent. We then managed to mechanize the duality principle and to build finite and infinite models, as well as desarguesian and non-desarguesian models. Using Coq helped us produce more precise proofs which handle all cases whereas in textbooks some very particular cases can sometimes be overlooked. Overall our Coq development of projective plane geometry amounts to 7.5K lines with about 200 definitions and lemmas.[5]

Our development makes use of a rather strong axiom, namely decidability of the incidence predicate. All subsequent properties are derived in an intuitionist logic from this axiom and those of projective plane geometry. It would be also interesting to perform our formalization using a purely constructive system of axioms as the ones proposed by Heyting and von Plato [14,28]. These systems of axioms are based on the apartness predicate which is the negation of the incidence predicate. It is easy to prove using classical logic that the axioms of a projective plane implies the axioms of Heyting. It would be also interesting to derive more theorems in a purely constructive framework.

In the future, we plan to carry on our investigations in two main directions. On the one hand, we expect to write a reliable algorithm for constraint solving in incidence geometry. It requires to specify projective plane geometry, which is what we achieve in this paper. The next step will be to certify that whenever the prover says three points are collinear (resp. non-collinear), we can build a proof at the specification level that these points are actually collinear (resp. non-collinear).

On the other hand, in a at least 3-dimensional setting, we shall study how to formally link together the axiom system based on the concept of ranks presented in [18] to a more traditional axiom system like the ones we presented in this paper. Moreover, formalizing in Coq some fully automated proof methods based on geometric algebras such as [29,17] would be an interesting challenge.

On the technical side, we also plan to study how our development can make use of first-order type classes [32] instead of modules and functors. We expect this new feature to improve the readability of the formal description by making implicit some technical details.

---

[5] The Coq development is available here:
http://coq.inria.fr/distrib/current/contribs/ProjectivePlaneGeometry.html

Acknowledgments. We would like to thank Loïc Pottier for providing the Coq tactic for solving systems of equations using Gröbner basis.

References

1. Bertot, Y., Castéran, P.: Interactive Theorem Proving and Program Development, Coq'Art: The Calculus of Inductive Constructions. Texts in Theoretical Computer Science. An EATCS Series. Springer, Heidelberg (2004)
2. Bezem, M., Hendriks, D.: On the Mechanization of the Proof of Hessenberg's Theorem in Coherent Logic. Journal of Automated Reasoning 40(1), 61–85 (2008)
3. Buekenhout, F. (ed.): Handbook of Incidence Geometry. North-Holland, Amsterdam (1995)
4. Cerroni, C.: Non-Desarguian Geometries and the Foundations of Geometry from David Hilbert to Ruth Moufang. Historia Mathematica 31(3), 320–336 (2004)
5. Chicli, L., Pottier, L., Simpson, D.: Mathematical quotients and quotient types in coq. In: Geuvers, H., Wiedijk, F. (eds.) TYPES 2002. LNCS, vol. 2646, pp. 95–107. Springer, Heidelberg (2003)
6. Chou, S.-C., Gao, X.-S., Zhang, J.-Z.: Machine Proofs in Geometry. Series on Applied Mathematics. World Scientific, Singapore (1994)
7. Chrząszcz, J.: Implementing Modules in the Coq System. In: Basin, D., Wolff, B. (eds.) TPHOLs 2003. LNCS, vol. 2758, pp. 270–286. Springer, Heidelberg (2003)
8. Coq development team, The: The Coq Proof Assistant Reference Manual, Version 8.0. LogiCal Project (2004)
9. Coxeter, H.S.M.: Projective Geometry. Springer, Heidelberg (1987)
10. Créci, J., Pottier, L.: Gb: une procédure de décision pour Coq. In: Actes JFLA 2004 (2004) (in french)
11. Dehlinger, C., Dufourd, J.-F., Schreck, P.: Higher-Order Intuitionistic Formalization and Proofs in Hilbert's Elementary Geometry. In: Richter-Gebert, J., Wang, D. (eds.) ADG 2000. LNCS (LNAI), vol. 2061, pp. 306–324. Springer, Heidelberg (2001)
12. Duprat, J.: Une axiomatique de la géométrie plane en Coq. In: Actes des JFLA 2008, pp. 123–136. INRIA (2008) (in french)
13. Guilhot, F.: Formalisation en Coq et visualisation d'un cours de géométrie pour le lycée. Revue des Sciences et Technologies de l'Information, Technique et Science Informatiques, Langages applicatifs 24, 1113–1138 (2005) (in french)
14. Heyting, A.: Axioms for intuitionistic plane affine geometry. In: Suppes, P., Henkin, A.T.L. (eds.) The Axiomatic Method, with Special Reference to Geometry and Physics, pp. 160–173. North-Holland, Amsterdam (1959)
15. Hoffmann, C.M., Joan-Arinyo, R.: Parametric Modeling. In: Handbook of Computer Aided Geometric Design, pp. 519–541. Elsevier, Amsterdam (2002)
16. Jermann, C., Trombettoni, G., Neveu, B., Mathis, P.: Decomposition of Geometric Constraint Systems: a Survey. International Journal of Computational Geometry and Application 16(5-6), 379–414 (2006)
17. Li, H., Wu, Y.: Automated Short Proof Generation for Projective Geometric Theorems with Cayley and Bracket Algebras: I. incidence geometry. J. Symb. Comput. 36(5), 717–762 (2003)
18. Magaud, N., Narboux, J., Schreck, P.: Formalizing Desargues' theorem in Coq using ranks. In: Proceedings of the ACM Symposium on Applied Computing, SAC 2009. ACM Press, New York (2009)

19. Meikle, L.I., Fleuriot, J.D.: Formalizing hilbert's grundlagen in isabelle/Isar. In: Basin, D., Wolff, B. (eds.) TPHOLs 2003. LNCS, vol. 2758, pp. 319–334. Springer, Heidelberg (2003)
20. Meikle, L.I., Fleuriot, J.D.: Mechanical Theorem Proving in Computational Geometry. In: Hong, H., Wang, D. (eds.) ADG 2004. LNCS (LNAI), vol. 3763, pp. 1–18. Springer, Heidelberg (2006)
21. Michelucci, D., Foufou, S., Lamarque, L., Schreck, P.: Geometric constraints solving: some tracks. In: SPM 2006: Proceedings of the 2006 ACM Symposium on Solid and Physical Modeling, pp. 185–196. ACM Press, New York (2006)
22. Miquel, A., Werner, B.: The Not So Simple Proof-Irrelevant Model of CC. In: Geuvers, H., Wiedijk, F. (eds.) TYPES 2002. LNCS, vol. 2646, pp. 240–258. Springer, Heidelberg (2003)
23. Moulton, F.R.: A Simple Non-Desarguesian Plane Geometry. Transactions of the American Mathematical Society 3(2), 192–195 (1902)
24. Narboux, J.: A Decision Procedure for Geometry in Coq. In: Slind, K., Bunker, A., Gopalakrishnan, G.C. (eds.) TPHOLs 2004. LNCS, vol. 3223, pp. 225–240. Springer, Heidelberg (2004)
25. Narboux, J.: Mechanical Theorem Proving in Tarski's Geometry. In: Botana, F., Recio, T. (eds.) ADG 2006. LNCS (LNAI), vol. 4869, pp. 139–156. Springer, Heidelberg (2007)
26. Paulson, L.C.: The Isabelle reference manual (2006)
27. Pichardie, D., Bertot, Y.: Formalizing Convex Hull Algorithms. In: Boulton, R.J., Jackson, P.B. (eds.) TPHOLs 2001. LNCS, vol. 2152, pp. 346–361. Springer, Heidelberg (2001)
28. von Plato, J.: The Axioms of Constructive Geometry. In: Annals of Pure and Applied Logic, vol. 76, pp. 169–200 (1995)
29. Richter-Gebert, J.: Mechanical Theorem Proving in Projective Geometry. Ann. Math. Artif. Intell. 13(1-2), 139–172 (1995)
30. Schreck, P.: Robustness in CAD Geometric Construction. In: 5th International Conference on Information Visualisation IV 2001, London, pp. 111–116 (July 2001)
31. Schwabhauser, W., Szmielew, W., Tarski, A.: Metamathematische Methoden in der Geometrie. Springer, Heidelberg (1983) (in german)
32. Sozeau, M., Oury, N.: First-Class Type Classes. In: Mohamed, O.A., Muñoz, C., Tahar, S. (eds.) TPHOLs 2008. LNCS, vol. 5170, pp. 278–293. Springer, Heidelberg (2008)
33. Tarski, A.: What is Elementary Geometry? In: Henkin, L., Suppes, P., Tarski, A. (eds.) The Axiomatic Method, with Special Reference to Geometry and Physics, pp. 16–29. North-Holland, Amsterdam (1959)

A   Proof of Equivalence of Axiom Systems

In this section, we provide the proof of the equivalence of the two axiom systems.

A.1   From Axiom 'Four Points' to Axiom 'Three Points' and Axiom 'Lower Dimension'

We first prove that each line contains at least three points:

$$\forall l : \text{Line}, \exists ABC : \text{Point}, \text{distinct3 } A \ B \ C \land A \in l \land B \in l \land C \in l.$$

We have as an assumption that there exist four points A, B, C and D with no three collinear. We have three cases to study depending on how many points are on line l: either two, one or zero points of these four points are on l.

- Two points are on l (say P and Q), two are not on l (say R and S).
  We build m which goes through R and S, it intersects l on a point (say X) which is different from P and Q. X has to be distinct from P (resp. Q), otherwise we would have P, R and S collinear (resp. Q, R and S collinear).
- One point is on l (say A) , the three other ones are not on l (say B, C and D).
  We have to build two more points on l. We proceed by creating lines going through points outside of l. We have to distinguish cases in order to avoid alignment issues.
- No point is on l, all four points (say A, B, C and D) are outside of l .
  We have to build three distinct points. We do the same reasoning steps, building lines from A, B, C and D.

All possible configuration for the 4 points can be captured by these three cases, sometimes via renaming of points. Details can be found in the formal Coq development.

Axiom 'Lower dimension' can be proved very easily:

$$\exists l_1 : \text{Line}, \exists l_2 : \text{Line}, l_1 \neq l_2.$$

We simply consider 2 lines l (which goes through A and B) and m (which goes through C and D). It is straightforward to show they are different: if they were not, then A, B, C and D would be collinear and this would contradict axiom 'Four points'.

A.2   From Axioms 'Three Points' and 'Lower Dimension' to Axiom 'Four Points'

We prove some preliminary lemmas: for any two distinct lines l_1 and l_2, each of them carrying at least three points (say M, N and O for l_1 and P, Q and R for l_2), we make a case distinction depending on where these points lie with respect to the intersection of l_1 and l_2. There are four cases to consider:

- One of the three points of l_1 (say M) and one of those of l_2 (say P, which is actually equal to M) are at the intersection of l_1 and l_2. Then the remaining points (M, O, Q and R) verify axiom 'Four points'. No three of them can be collinear otherwise we would have $l_1 = l_2$.
- One point of l_1 (say M) is at the intersection of l_1 and l_2. Then points M, O, Q and R verify axiom 'Four points'.
- One point of l_2 (say P) is at the intersection of l_1 and l_2. Exchanging l_1 and l_2 in the previous lemma solves the case.
- No point of l_1 and l_2 is at the intersection. Then points M, O, Q and R also verify axiom 'Four points'.

Axiom 'Four points' is then proved by first making two lines l_1 and l_2 explicit (through axiom 'Lower dimension'), then considering three distinct points on each line (through axiom 'Three points'). The four lemmas allow to prove the existence of four points in the various possible configurations depending on which points (if any) lie at the intersection of l_1 and l_2.

B Decidability Proofs

From the axiom system ProjectivePlane (see Fig. 2) and a decidability axiom about incidence, namely

$$\forall A : \text{Point}, l : \text{Line}, A \in l \vee \neg A \in l,$$

we can derive proofs of decidability of point equality as well as line equality. Both theorems can be proved independently, in a intuitionist way (none of them require the use of classical logic).

B.1 Line Equality

Given any two lines l_1 and l_2, they either are equal or not:

$$\forall l_1\, l_2 : \text{Line}, l_1 = l_2 \vee l_1 \neq l_2.$$

From axiom 'Three points', we know there exists three distinct points M, N and P on l_1. We then proceed by case analysis depending on whether M and N are on l_2.

$M \in l_2$:
    $N \in l_2$: $l_1 = l_2$ because of the Uniqueness Axiom and the fact that $M \neq N$.
    $N \notin l_2$: l_1 and l_2 are different because N is on l_1 and not on l_2.
$M \notin l_2$: l_1 and l_2 are different because M is on l_1 and not on l_2.

B.2 Point Equality

Given any two points A and B, they either are equal or not:

$$\forall A\, B : \text{Point}, A = B \vee A \neq B.$$

We first prove an auxiliary lemma:

$$\forall A\, B : \text{Point}, \forall d : \text{Line}, A \notin d \Rightarrow B \notin d \Rightarrow A = B \vee A \neq B.$$

From axiom 'Three points', we know there exists three distinct points M, N and P incident to d. We build two lines $l_1 = (AM)$ and $l_2 = (AN)$. These two lines are different because N and M are distinct and A is not incident to d.

$B \in l_1$:
    $B \in l_2$: $A = B$ from the Uniqueness Axiom and the fact that $l_1 \neq l_2$.

$B \notin l_2$: $A \neq B$ because A is incident to l_2 and B is not.
$B \notin l_1$: $A \neq B$ because A is incident to l_1 and B is not.

The main theorem can now be proved: from axiom 'Lower dimension', there exists two distinct lines Δ_0 and Δ_1. We proceed by case analysis on whether A and B belong to Δ_0 and Δ_1.

$A \in \Delta_0$:
   $B \in \Delta_0$:
      $A \in \Delta_1$:
         $B \in \Delta_1$: $A = B$ from the Uniqueness Axiom and the fact that
            $\Delta_0 \neq \Delta_1$.
         $B \notin \Delta_1$: $A \neq B$, because A is incident to Δ_1 and B is not.
      $A \notin \Delta_1$:
         $B \in \Delta_1$: $A \neq B$, because A is not incident to Δ_1 and B is.
         $B \notin \Delta_1$: We apply the previous lemma with $d = \Delta_1$.
   $B \notin \Delta_0$: $A \neq B$, because A is incident to Δ_0 and B is not.
$A \notin \Delta_0$:
   $B \in \Delta_0$: $A \neq B$, because A is not incident to Δ_0 and B is.
   $B \notin \Delta_0$: We apply the previous lemma with $d = \Delta_0$.

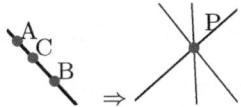

Fig. 6. Proving the dual of axiom 'Three points'

C   Proof of the Principle of Duality

In this section, we show that the duality principle is correct. As stated before the proof of most axioms is straightforward, hence we only prove the dual of axiom 'Three points'. We need to prove that:

$$\forall P, \exists l_1 l_2 l_3, P \in l_1 \wedge P \in l_2 \wedge P \in l_3.$$

First, we prove the following two lemmas:

$$\forall Pl, P \notin l \Rightarrow \exists l_1 l_2 l_3, P \in l_1 \wedge P \in l_2 \wedge P \in l_3$$

and

$$\forall l_1 l_2, \exists P, P \notin l_1 \wedge P \notin l_2.$$

Proof of the first lemma: let's take three distinct points A, B and C on l using axiom 'Three points'. Then we can build the lines (PA), (PB) and (PC). Those lines are distinct because otherwise using the uniqueness axiom we could prove that A, B and C are not distinct.

Proof of the second lemma: If $l_1 = l_2$ we need to build a point not on l_1. From axiom 'Lower dimension', we know there are two lines. From axiom 'Three points' we can conclude because we know there are at least three points on each line.

Otherwise $l_1 \neq l_2$. Let's call C the intersection of l_1 and l_2. Then, we can build two points P_1 and P_2 on l_1 and l_2 respectively which are different from C. We know that $P_1 \neq P_2$ because otherwise $l_1 = l_2$. Let d be the line through P_1 and P_2. We can build a third point Q on d. Q is neither on l_1 nor on l_2. This concludes the lemma.

Finally, we can prove the dual of axiom 'Three points'. We build two lines l_1 and l_2 using axiom 'Lower dimension'. Then we perform case distinction on $P \in l_1$ and $P \in l_2$. If $P \in l_1 \wedge P \in l_2$ we use the second lemma. Otherwise $P \notin l_1 \vee P \notin l_2$. In both cases, we can use the first lemma.

E   Lines as Set of Points

In our development, we consider two basic notions: points and lines. Lines can actually be viewed as sets of points. With this representation, for any lines l_1 and l_2 we can build a bijection from l_1 to l_2.

We first define the set of points corresponding to a given line l, it consists of all the points of the plane which are incident to l.

```
Definition line_as_set_of_points (l:Line):= {X:Point | Incid X l}.
```

From this definition, we want to prove the following theorem:

```
Theorem line_set_of_points : forall l1 l2:line,
exists f:(line_as_set_of_points l1) -> (line_as_set_of_points l2),
        bijective f.
```

It states there exists a bijective function f from l_1 to l_2 when these lines are viewed as sets of points. We build a constructive proof of this existential formula, which requires to make explicit the function f and then check whether it is actually a bijection, i.e. verifies the one-to-one and onto properties.

The proof proceeds as follows:

First of all, one can safely assume that l_1 and l_2 are different. If not, then the identity function works just fine. The first step of the proof is to write a function which, given two lines l_1 and l_2 computes a point P which belongs neither to l_1 nor to l_2.

```
Lemma outsider : forall l1 l2: Line,
             {P:Point | ~Incid P l1/\~Incid P l2}.
```

We now explicitly construct the function f as shown on Fig. 7. Given a point A_1 of l_1, we can build a line (say Δ) going through A_1 and P. Lines Δ and l_2 intersect in a point A_2. We define f such that $f(A_1) = A_2$. It remains to prove

D Incidence Relation of PG(2,5)

Table 1. The incidence relation of PG(2,5). Each column lists the lines incident to the given point.

P_{30}	P_{29}	P_{28}	P_{27}	P_{26}	P_{25}	P_{24}	P_{23}	P_{22}	P_{21}	P_{20}	P_{19}	P_{18}	P_{17}	P_{16}	P_{15}	P_{14}	P_{13}	P_{12}	P_{11}	P_{10}	P_9	P_8	P_7	P_6	P_5	P_4	P_3	P_2	P_1	P_0
1	2	3	4	5	6	7	8	9	10	11	12	13	14	15	16	17	18	19	20	21	22	23	24	25	26	27	28	29	30	0
2	3	4	5	6	7	8	9	10	11	12	13	14	15	16	17	18	19	20	21	22	23	24	25	26	27	28	29	30	0	1
4	5	6	7	8	9	10	11	12	13	14	15	16	17	18	19	20	21	22	23	24	25	26	27	28	29	30	0	1	2	3
9	10	11	12	13	14	15	16	17	18	19	20	21	22	23	24	25	26	27	28	29	30	0	1	2	3	4	5	6	7	8
13	14	15	16	17	18	19	20	21	22	23	24	25	26	27	28	29	30	0	1	2	3	4	5	6	7	8	9	10	11	12
19	20	21	22	23	24	25	26	27	28	29	30	0	1	2	3	4	5	6	7	8	9	10	11	12	13	14	15	16	17	18

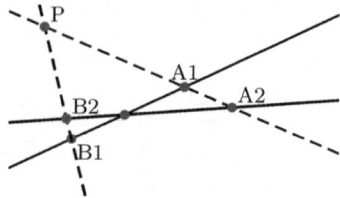

Fig. 7. Building a bijection between l_1 and l_2

that this function is actually bijective. Proving that this function is one-to-one requires to assume proof irrelevance [22]. Proof irrelevance expresses that proofs of the same formula are equal. It allows us to show existential propositions with the same type are equal regardless of the proof terms proving the formulas. Proving the onto property requires to apply the construction process of f in the reverse order going from line l_2 to line l_1.

Linear Programming for Bernstein Based Solvers

Dominique Michelucci and Christoph Fünfzig

LE2I, UMR CNRS 5158, 9 av Alain Savary, BP 47870, 21078 Dijon cedex, France
Dominique.Michelucci@u-bourgogne.fr

Abstract. Some interval Newton solvers rely on tensorial Bernstein bases to compute sharp enclosures of multivariate polynomials on the unit hypercube. These solvers compute all coefficients with respect to tensorial Bernstein bases. Unfortunately, polynomials become exponential size in tensorial Bernstein bases. This article gives the first polynomial time method to solve this issue. A polynomial number of relevant Bernstein polynomials is selected. The non-negativity of each of these Bernstein polynomials gives a linear inequality in a space connected to the monomials of the canonical tensorial basis. We resort to linear programming on the resulting *Bernstein polytope* to compute range bounds of a polynomial or bounds of the zero set.

Keywords: subdivision solver, tensorial bernstein basis, Bernstein polytope, geometric constraint solving.

1 Introduction

Especially in 3D, geometric constraint solving eventually requires solving systems of non-linear, typically algebraic equations. Usually, irreducible systems are solved with numerical methods, for instance homotopy [12,4,17], Newton-Raphson iterations, interval Newton methods [9], and Bernstein-based solvers.

Computer Graphics, CAD-CAM, and some people in numerical analysis, use properties of tensorial Bernstein bases (TBB) and Bernstein based solvers, for computing intersections between algebraic non-linear surfaces and curves, and for numerically solving systems of polynomial equations [8,13,16,5,15,11]. TBB provide sharp enclosures of multivariate polynomials over a box, *i.e.*, a cartesian product of intervals. The range of a multivariate polynomial $p(x)$, $x = (x_1, \ldots, x_n)$ over the unit box $x \in [0,1]^n$ is the interval given by the smallest and the greatest coefficients of the polynomial $p(x)$ expressed in the TBB. This property and the de Casteljau algorithm or other subdivision methods are used in Computer Graphics and CAD-CAM to compute tight covers of implicit algebraic curves and surfaces [10].

However, polynomials become exponential size in the TBB. For instance, the monomial 1 is written $(B_0^{(d_1)}(x_1) + \ldots + B_{d_1}^{(d_1)}(x_1)) \cdot \ldots \cdot (B_0^{(d_n)}(x_n) + \ldots + B_{d_n}^{(d_n)}(x_n))$. Even a linear polynomial $p(x_1, \ldots, x_n)$ has an exponential number 2^n of coefficients in the TBB, while it has linear size $O(n)$ in the commonly used canonical basis. A quadratic polynomial $p(x_1, \ldots, x_n)$ has exponential size 3^n in

T. Sturm and C. Zengler (Eds.): ADG 2008, LNAI 6301, pp. 163–178, 2011.
© Springer-Verlag Berlin Heidelberg 2011

the TBB, while it has a quadratic number $O(n^2)$ of coefficients in the canonical basis with monomials x_i^2, $x_i x_j$, x_i, and 1.

This feature makes current Bernstein based solvers impracticable for systems with more than $n = 6$ or 7 unknowns. Geometric constraints, especially in 3D, often yield big irreducible systems. For instance, the regular icosahedron and non-regular ones (20 triangular faces, 12 vertices, 30 edges) can be specified, up to location and orientation in 3D space, by the length of their edges. Similar for their duals, the regular dodecahedron, or non-regular ones (12 pentagonal faces, 20 vertices, 30 edges) can be specified by the length of their edges and by coplanarity conditions for each of their faces. These systems of equations are quadratic, *i.e.*, the degree of their monomials is at most 2 since equations are: $a_k^2 + b_k^2 + c_k^2 = 1$, $a_k x_i + b_k y_i + c_k z_i + d_k = 0$, $(x_i - x_j)^2 + (y_i - y_j)^2 + (z_i - z_j)^2 = D_{ij}^2$, where (x_i, y_i, z_i) are the coordinates of vertex i, $a_k x + b_k y + c_k z + d_k = 0$ is the equation of the plane of face k, and D_{ij} is the length of edge ij. For a well-constrained system, three points are arbitrarily fixed: a vertex is fixed at the origin, one of its neighbor on the x axis, and another one is fixed on the Oxy plane. These systems are huge, they are roughly irreducible, and they can not be solved with current TBB solvers due to the exponential size of the representation of their polynomials.

This article proposes the first polynomial time algorithm to overcome this difficulty with TBB. The solver in [15] does not use the TBB but resorts to the simplicial Bernstein bases, which have polynomial cardinality.

The main idea is to make explicit the Bernstein polytope. While current TBB solvers implicitly use this kind of polytope and consider it as the convex hull of its vertices, the method in this article considers it as the intersection of a set of halfspaces in a space connected to the monomials of the canonical tensorial basis: the non-negativity of some Bernstein polynomials provide linear inequalities for each non-linear monomial. It turns out that the Bernstein polytope has an exponential number of vertices, but it has only a polynomial number of bounding hyperplanes. Thus linear programming (LP) algorithms, say the simplex algorithm, can be used to compute the vertex of the Bernstein polytope, which minimizes or maximizes the linear objective function corresponding to a polynomial. It is known that the simplex method is not polynomial time in the worst case thus it is better to invoke a polynomial time method like the ellipsoid method or an interior point method from a theoretical point of view. However in practice the simplex algorithm is very competitive.

Since a lot of systems of geometric constraints yield to systems of quadratic equations (like the examples above), and since all algebraic systems can be reduced with polynomial overhead to quadratic systems using auxilliary equations and variables and using iterated squaring, this paper only considers systems of quadratic equations for simplicity. The main idea of the algorithm is the following: define the Bernstein polytope through its hyperplanes rather than as the convex hull of its vertices. This polytope definition can straightforwardly be extended to systems with higher degree.

Section 2 reminds standard notations. Section 3 defines the Bernstein polytope and its bounding hyperplanes. Sections 4.1 and 4.2 explain how to compute a range of a multivariate polynomial and reducing a domain while preserving its roots reduces to linear programming problems, considering the Bernstein polytope. Computing tight range bounds of multivariate polynomials can be used in interval Newton solvers, the principle of which is presented in Sect. 5. Actually, the Bernstein polytope can be used to propose a new kind of solver, which does not use Newton's method. It is presented in Sect. 6. Some technical problems, scaling and inaccuracy, are succinctly mentioned in Sect. 7. Our first solver implementation bypasses the inaccuracy issue using rational arithmetic; we then implemented the first robust floating-point CPU variant of this solver (Sect. 7.4). Section 7.3 mentions some theorems used by solvers to certify that a box contains no root, or at least one root, or a unique regular root. Section 8 concludes and lists future works.

2   Tensorial Bernstein Bases, Definitions, Main Properties

The $d+1$ Bernstein polynomials $B_i^{(d)}$ of degree d, also written B_i for fixed d, are a basis of degree-d polynomials

$$B_i^{(d)}(x) = \binom{d}{i} x^i (1-x)^{d-i}, \ i = 0, \ldots, d,$$

where the binomial coefficient $C(i, d) = \binom{d}{i}$ is the number of i-subsets of a d-set.

The conversion with the canonical basis $(x^0, x^1, \ldots, x^d)$ is a linear mapping. Classical formulas are

$$x^k = (1/C(k, d)) \sum_{i=k}^{d} C(k, i) \, B_i^{(d)}(x)$$

$$x^1 = (1/d) \sum_{i=0}^{d} i \, B_i^{(d)}(x)$$

$$x^0 = 1 = \sum_{i=0}^{d} B_i^{(d)}(x).$$

The main properties are that their sum equals 1, and every $B_i^{(d)}(x)$ is positive for $x \in [0, 1]$.

It means that for $0 \le x \le 1$, $p(x) = \sum p_i B_i(x)$ is a linear convex combination of the coefficients p_i. For a polynomial p with $p_i \in \mathbb{R}$, $p(x)$, $x \in [0, 1]$ lies in the interval $[\min p_i, \max p_i]$. This enclosure is tight, and the minimum or maximum bound is exact if it is attained for $i = 0$ or $i = d$. If the p_i lie in 2D (or 3D), $p(x)$ describes a 2D (or 3D) Bézier curve, and the arc $p(x)$, $x \in [0, 1]$ lies inside the convex hull of its so called control points p_i.

For an example, let $p(x)$ be a polynomial in $x \in \mathbb{R}$. Since $x = 0\,B_0(x) + 1/d\,B_1(x) + 2/d\,B_2(x) + \ldots + d/d\,B_d(x)$, the polynomial curve $(x, y = p(x))$ with $x \in [0, 1]$ lies in the convex hull of its control points $(i/d, p_i)$, where $p(x) = \sum_i p_i B_i(x)$.

In constrast to coefficients in the canonical basis $(1, x, x^2, \ldots, x^d)$, control points depend on the interval for x. The classical de Casteljau method provides the control points of $p(x)$, $x \in [0, t]$ and of $p(x)$, $x \in [t, 1]$.

For multivariate polynomials, the TBB is the tensorial product

$$(B_0^{(d_1)}(x_1), \ldots, B_{d_1}^{(d_1)}(x_1)) \cdot (B_0^{(d_2)}(x_2), \ldots, B_{d_2}^{(d_2)}(x_2)) \cdot \ldots.$$

The convex hull properties and the de Casteljau method extend to the TBB, which provide sharp enclosure of multivariate polynomials $p(x)$, $x \in [0, 1]^n$. For this reason, TBB are routinely used in CAD-CAM and Computer Graphics for computing tight covers, *e.g.*, voxelizations, of implicit algebraic curves or surfaces [10] in low dimension (2D, 3D, 4D).

3   Definition of the Bernstein Polytope

Each monomial x_i, $x_i x_j$, x_i^2 with total degree 1 or 2 is attached to a variable of a linear programming problem. Non-linear dependences between monomials x_i and x_i^2, or between monomials x_i, x_j, $x_i x_j$, are represented by the Bernstein polytope, through linear inequalities constraining corresponding LP variables.

3.1   Univariate Polynomials

For univariate polynomials of degree d, the Bernstein polytope is a convex polyhedron which encloses the arc of the curve $(x, x^2, \ldots, x^d)$ in $\mathbb{R}^d$ with $x \in [0, 1]$. Its hyperplanes and halfspaces are given by $B_i^{(d)}(x) \geq 0$, $i = 0, \ldots, d$.

For degree $d = 2$, the Bernstein polytope is a triangle, in Fig. 1; x and y are the LP variables representing monomials x and x^2; counting multiplicities, each triangle side meets the curve $(x, y = x^2)$, $0 \leq x \leq 1$, in two points. For degree $d = 3$, the Bernstein polytope is a tetrahedron, in Fig. 2; x, y, z are the LP variables representing monomials x, x^2, x^3; counting multiplicities, each plane meets the curve $(x, y = x^2, z = x^3)$, $0 \leq x \leq 1$ in three points. The extension to higher degrees is easy.

3.2   Multivariate Polynomials

We extend the Bernstein polytope to multivariate polynomials as follows. The inequalities for multivariate polynomials are obtained as the relevant products of the inequalities for univariate polynomials. We consider only quadratic polynomials with monomials $x_i, x_i^2, x_i x_j, x_j^2$. The hyperplanes for x_i, x_i^2 have been given above. So consider now the non-linear dependences between monomials $x_i, x_j, x_i x_j$, renamed $x, y, z = xy$ in the following for a more intuitive notation. As usual, all variables have values from the unit interval $[0, 1]$. The surface

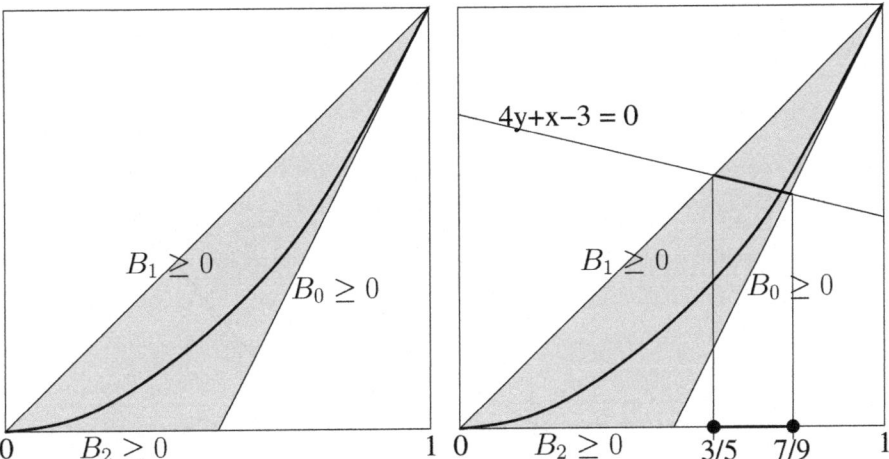

Fig. 1. Left: The Bernstein polytope encloses the curve $(x, y = x^2)$ for $(x, y) \in [0, 1]^2$. Its limiting sides are: $B_0(x) = (1-x)^2 = y - 2x + 1 \geq 0$, $B_1(x) = 2x(1-x) = 2x - 2y \geq 0$, $B_2(x) = x^2 = y \geq 0$. Right: solving $4x^2 + x - 3 = 0$ with $x \in [0, 1]$ is equivalent to intersecting the line $4y + x - 3 = 0$ with the curve (x, x^2). By linear programming, the x interval is reduced from $[0, 1]$ to $[3/5, 7/9]$.

patch $(x, y, z = xy)$ is enclosed in a convex polyhedron, shown in Fig. 3, whose halfspaces are

$$
\begin{aligned}
B_0^{(1)}(x) B_0^{(1)}(y) = (1-x)(1-y) \geq 0 &\Rightarrow 1 - x - y + z \geq 0 \\
B_0^{(1)}(x) B_1^{(1)}(y) = (1-x)y \geq 0 &\Rightarrow y - z \geq 0 \\
B_1^{(1)}(x) B_0^{(1)}(y) = x(1-y) \geq 0 &\Rightarrow x - z \geq 0 \\
B_1^{(1)}(x) B_1^{(1)}(y) = xy \geq 0 &\Rightarrow z \geq 0.
\end{aligned}
$$

This tetrahedon is the convex hull of the patch, thus it is optimal. Each of these non-linear inequalities $B_i^{(1)}(x) B_j^{(1)}(y) \geq 0$ in x and y gives a linear inequality in the LP variables x, y, z. For a quadratic system in n unknowns $x_1, \ldots, x_n$, their number $O(n^2)$ is polynomial. It is of the same order as an arbitrary quadratic polynomial in the canonical basis. The extension to higher degrees is easy but left to the reader for conciseness.

4   Linear Programming

The method resorts to linear programming (LP) to bypass the problem due to the exponential cardinality of the TBB. This section shows how to compute a range for a polynomial $p(x)$, $x = (x_1, \ldots, x_n)$ over the unit hypercube, and how it reduces to solving a linear programming problem.

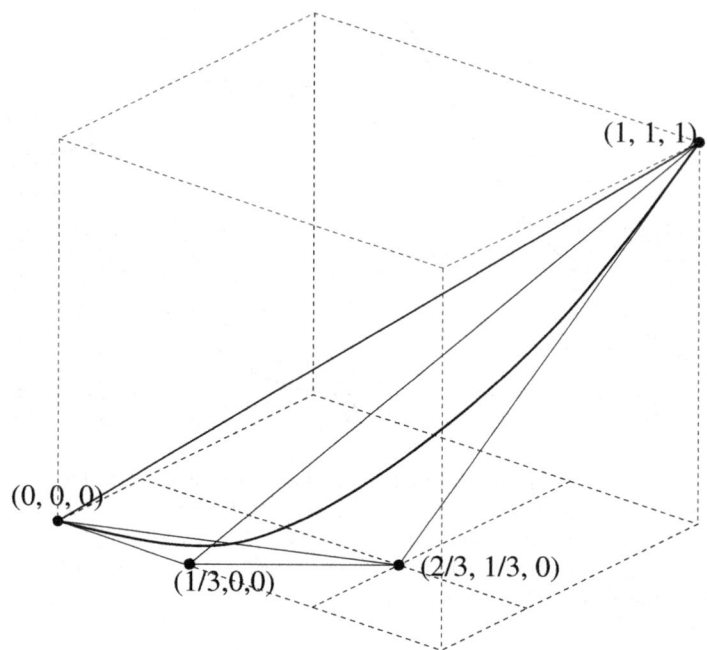

Fig. 2. The Bernstein polytope, a tetrahedron, enclosing the curve $(x, y = x^2, z = x^3)$ with $x \in [0, 1]$. Its vertices are $v_0 = (0, 0, 0)$, $v_1 = (1/3, 0, 0)$, $v_2 = (2/3, 1/3, 0)$ and $v_3 = (1, 1, 1)$. v_0 lies on $B_1 = B_2 = B_3 = 0$, v_1 on $B_0 = B_2 = B_3 = 0$, etc. $B_0(x) = (1 - x)^3 \geq 0 \Rightarrow 1 - 3x + 3y - z \geq 0$, $B_1(x) = 3x(1 - x)^2 \geq 0 \Rightarrow 3x - 6y + 3z \geq 0$, $B_2(x) = 3x^2(1 - x) \geq 0 \Rightarrow 3y - 3z \geq 0$, $B_3(x) = x^3 \geq 0 \Rightarrow 3z \geq 0$.

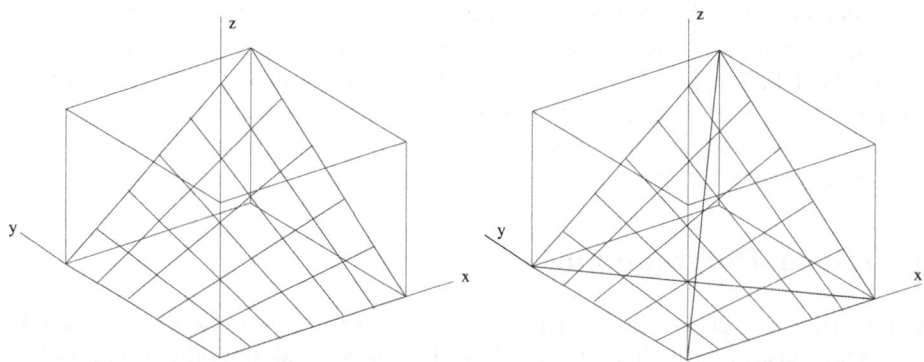

Fig. 3. The Bernstein polytope enclosing the surface patch $(x, y, z = xy)$. Inequalities of delimiting planes are $B_i(x)B_j(y) \geq 0$, where $B_0(t) = 1 - t$, $B_1(t) = t$ and $i = 0, 1$.

4.1   Range Bound for a Polynomial

The method for computing a range bound of a polynomial is illustrated with a simple example. To compute a lower and an upper bound of the polynomial $p(x) = 4x^2 + x - 3$, for $x \in [0, 1]$, minimize, and maximize, the linear objective function: $4y + x - 3$ on the Bernstein polytope (the triangle in Fig. 1) enclosing the curve $(x, y = x^2)$, $x \in [0, 1]$. It is a LP problem after replacing x^2 with y. From left to right, the LP tableau of the initial problems, the LP tableau for the minimum, the LP tableau for the maximum are given:

$$
\begin{array}{lll}
\min, \max : p = 4y + x - 3 & \min p = -3 + x + 4y & \max p = 2 - 5B_0 - 9/2B_1 \\
0 \le B_0 = y - 2x + 1 & B_0 = 1 - 2x + y & x = 1 - B_0 - B_1/2 \\
0 \le B_1 = -2y + 2x & B_1 = 2x - 2y & y = 1 - B_0 - B_1 \\
0 \le B_2 = y, & B_2 = y, & B_2 = 1 - B_0 - B_1.
\end{array}
$$

The simplex method due to Dantzig performs Gauss pivoting operations on the rows of the initial tableau to reach the two last tableaux, which exhibit the minimum and the maximum. Variables on the left side of the tableaux are basic variables, variables on the right side are non-basic variables with values 0, the standard convention in linear programming. Let us comment on the tableau for the maximum. In $\max p = 2 - 5B_0 - 9/2B_1$, the value of p can not be greater than 2 because non-basic variables B_0 and B_1 have values 0, and increasing their values can only decrease p due to their negative coefficients $-5B_0 - 9/2B_1$ in the objective function. The same kind of comments apply to the tableau for the minimum. Thus the polynomial $p(x \in [0, 1])$ lies in the interval $[-3, 2]$. The minimum occurs at $x = 0$ (x is a non-basic variable for the minimum tableau) so it is exact. The maximum occurs at $x = 1$ (x is a basic variable in the line $x = 1 - B_0 - B_1/2$ of the rightmost tableau) so it is exact too.

Observe that at vertex $v_0 = (0, 0)$ where $B_1 = B_2 = 0$, the polynomial value is $p_0 = p(0) = -3$; at vertex $v_1 = (1/2, 0)$ where $B_0 = B_2 = 0$, the polynomial value is $p_1 = p(1/2) = -3/2$; at vertex $v_2 = (1, 1)$ where $B_0 = B_1 = 0$, the polynomial value is $p_2 = p(1) = 2$. These values p_0, p_1, p_2 are the coefficients in the Bernstein basis of $p(x)$: $p(x) = p_0B_0(x) + p_1B_1(x) + p_2B_2(x)$.

This property trivially holds for all univariate polynomials by definition of the Bernstein polytope.

Remark 1. It is possible to drop the coefficients in the inequalities $B_i^{(d)}(x) \ge 0$ of the halfspaces of the Bernstein polytope. It does not modify the Bernstein polytope and does not modify the results of range computations, nor the results of the domain reduction (Sect. 4.2).

Remark 2. The Bernstein polytope for univariate polynomials can be tightened, e.g., with $B_1^{(2)}(x_i) \le 1/2$ as in Fig. 4. The number of hyperplanes is still polynomial, and tighter ranges are obtained. It can also improve the reduction of domains. Since inequalities for multivariate polynomials are just products of inequalities for univariate ones, the Bernstein polytope for multivariate polynomials (with degree greater than 2) can also be tightened. Adding halfspaces is not possible with current TBB solvers that use the primal definition.

Remark 3. A forerunner of this approach is Olivier Beaumont [1], who used Chebychev polynomials and LP for enclosing multivariate polynomials in his PhD. In the univariate case, Chebychev inequalities are obtained as follows: the monomial x^d is interpolated in the d Gauss points with a degree $d-1$ polynomial $T(x)$, then the error is bounded, which gives inequalities $b_0 \leq x^d - T(x) \leq b_1$. Actually, any interpolation scheme using its own interpolation points, *e.g.*, the minimax polynomials, gives inequalities. Inequalities in the multivariate case $(x_1, x_2, \ldots, x_n)$ are again given by relevant products of univariate inequalities. Chebychev polynomials or the minimax polynomials provide other inequalities and other halfspaces, which can be used instead or together with the Bernstein polytope.

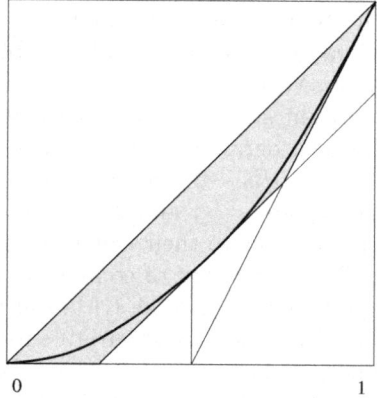

Fig. 4. Tightened Bernstein polytope, *e.g.*, using the inequality $x - y \leq 1/4$

4.2   Domain Reduction

This section shows how the solver reduces intervals or boxes, preserving the contained roots, for the simple equation $4x^2 + x - 3 = 0$ for $x \in [0, 1]$. Solving is equivalent to finding the intersection points between the line $4y + x - 3 = 0$ and the curve $(x, y = x^2)$. This curve is enclosed in its Bernstein polytope, the triangle of Fig. 1. Intersecting the line and the triangle, *i.e.*, finding the minimum and maximum value of x, will reduce the interval for x. It is the same LP problem as above except this time we minimize and maximize x. The LP tableaux are

$$\begin{array}{lll}
\min, \max : x & \min x = 3/5 + 2/5 B_1 & \max x = 7/9 - 4/9 B_0 \\
0 \leq B_0 = y - 2x + 1 & y = 3/5 - 1/10 B_1 & y = 5/9 + 1/9 B_0 \\
0 \leq B_1 = -2y + 2x & B_0 = 2/5 - 9/10 B_1 & B_1 = 4/9 - 10/9 B_0 \\
0 \leq B_2 = y, & B_2 = 3/5 - 1/10 B_1, & B_2 = 5/9 + 1/9 B_0.
\end{array}$$

Thus the interval $[0,1]$ for x is reduced to $[3/5, 7/9]$, and no root is lost. To further reduce this interval, use the scaling in Sect. 7.1, which maps $x \in [3/5, 7/9]$ to $X \in [0,1]$: $x = 3/5 + (7/9 - 3/5)X = b + aX$, and the equation in X is $4a^2X^2 + (8ab+b)X + (b-3) = 0$. Convergence around a regular root is quadratic like Newton's method but the convergence rate is not discussed in detail here.

If the line does not cut the Bernstein triangle, more generally, if the LP problem is not feasible then it proves that the domain contains no root.

5   Use in Interval Newton Solvers

TBB-based solvers are interval Newton solvers, which rely on the TBB properties to compute tight ranges of multivariate polynomials.

This section presents the principle of interval Newton methods, which isolates real roots of a well-constrained system $f(x) = 0$, $x \in \mathbb{R}^n$ and $f : \mathbb{R}^n \to \mathbb{R}^n$, inside a given initial box $B \subset \mathbb{R}^n$. Push B on a stack of boxes to be studied. First try to reduce B: compute with some interval method a range B' (an enclosing box) of $N(B)$, where $N(x) = x - f(x)M$, where M is the inverse of the jacobian of f at the centre of the box B. As usual, a floating point approximation of the inverse is sufficient, and an LU decomposition can be used instead of explicitly computing the inverse matrix. Roots inside B are located in $B \cap B'$. If $B \cap B'$ is empty, B contains no root. Otherwise, if $B \cap B'$ is significantly smaller than B, try to reduce $B \cap B'$ again, or, if some Kantorovich test guarantees that there is a unique root inside and that Newton iterations are going to converge (Sect. 7.3), apply the classical Newton method to the center of $B \cap B'$, and add the resulting root to a list of solutions. If $B \cap B'$ is not significantly smaller than B, bissect $B \cap B'$ for instance along its longest side, or the side which reduced the least in the current iteration, and push the two halves on the stack. Actually, a set of residual boxes is typically handled: a box is residual when the box is small and can no more be divided because of the finite precision of floating point arithmetic, but the method can not decide on the status of the box, for instance, it contains a singular root or very close regular roots.

The main difficulty in this algorithm is to compute a tight range bound of $N(B)$ and this is one topic of this article. Computing an ϵ-approximation of the exact range is NP-hard. Thus researchers in the interval analysis community have proposed several methods to compute in polynomial time a superset of the exact range with some trade offs between time complexity and accuracy. It turns out that TBB provide sharp range bounds.

If all equations are quadratic, it is composed of n quadratic polynomials $P_i(x)$, and it is easy to symbolically compute $N(x)$. Each polynomial is defined by $O(n^2)$ coefficients, represented by floating point values or better by intervals with floating point bounds. We can also apply the scaling (Sect. 7.1) in polynomial time so that the studied box B is $[0,1]^n$. The main problem is then to compute a range bound of a quadratic polynomial $p(x)$ with $x \in [0,1]^n$. For this, the method based on TBB and LP is explained in Sect. 4.1.

6   New Solver

The principle of the Bernstein polytope can be used to compute tight range bounds of multivariate polynomials and applied to classical interval Newton methods. However, the Bernstein polytope or tightened Bernstein polytopes make possible new solvers, which no more refer to Newton's method.

In this new method, the Bernstein polytope enclosing the quadratic algebraic patch $(x_1, \ldots, x_n, x_1^2, \ldots, x_n^2, x_1 x_2, \ldots, x_{n-1} x_n)$ is defined as before. Moreover, all equations of $f(x) = 0$ are translated into n linear constraints in the LP variables. Quadratic inequalities can also be translated into linear inequalities in the LP variables: this approach deals very easily with inequalities, in constrast to other solvers like homotopy solvers. Then $2n$ linear optimization problems are solved: minimize x_i for $i = 1, \ldots, n$ and maximize x_i for $i = 1, \ldots, n$.

Figure 1, right, shows this method applied to the equation $4x^2 + x - 3 = 0$. Let y be the LP variable representing x^2, corresponding to variable x. The intersection of the convex polygon and the line $4y + x - 3 = 0$ gives an interval $[3/5, 7/9]$ for x, which encloses the root of the equation: $4x^2 + x - 3 = 0$. This interval is then mapped to $[0, 1]$: $x = 3/5 + (7/9 - 3/5)X$, $X \in [0, 1]$ using the scaling in Sect. 7.1. The same method is then applied to the resulting equation in X. The convergence is quadratic if there is only a regular root. When the box is not significantly reduced, for instance for $x^2 - x = 0$, a bisection is performed as usual. Empirically, almost all bisections separate roots. Note that bisections are the only way to separate roots, domain reduction can not.

An advantage of this solver is that preconditioning the system, *i.e.*, multiplying equations with the inverse of the jacobian at the center of the box, is not necessary. All the work is performed by the simplex method.

Indeed, current TBB solvers often need and use some specific procedure to detect as early as possible that the studied box contains no root [13,11] (Sect. 7.3) in order to avoid an exponential number of bisections, when separating two close and locally parallel curves in 2D. A recent article [11] proposed a procedure which also takes into account inequalities $g_i(x) \leq 0$ in the system $f(x) = 0$, $g(x) \leq 0$. Its principle is to search with linear programming a polynomial $h(x) = \sum_j \alpha_j f_j(x) + \sum \beta_i g_i(x)$ with $\alpha_j \in \mathbb{R}$ and $\beta_i \geq 0$ such that $h(x)$ is always greater than 1 in the studied box, *i.e.*, its smallest coefficient in the TBB is 1. If such an h exists then the studied box contains no root. It turns out that this new solver detects early that boxes contain no root without any specific procedures, *i.e.*, the LP problem is not feasible in this case.

7   Technicalities

7.1   Scaling

After reduction, the reduced box is no more $[0, 1]^n$. A scaling maps the box $[u, v]$ with $u_i \leq v_i$ to the unit hypercube $[0, 1]^n$. Define $x_i = u_i + (v_i - u_i)X_i$ with $w_i = v_i - u_i$, $X \in [0, 1]^n$. Then $x_i^2 = w_i^2 X_i^2 + 2u_i w_i X_i + u_i^2$,

$x_i x_j = w_i w_j X_i X_j + u_i w_j X_j + u_j w_i X_i + u_i u_j$. Scaling is a linear mapping in the space of the LP variables.

Another possibility is to scale the Bernstein polytope. The equalities and inequalities of the system $f(x) = 0, g(x) \leq 0$ are left unchanged by this. As usual in the LP problem, the monomial x_i^2 is represented by some LP variable q_i, and the monomial $x_i x_j$ by some LP variable x_{ij}. For a box $x_i = [u_i, v_i]$ with $w_i = v_i - u_i$, the hyperplanes of the Bernstein polytope are changed as follows

$$B_0^{(2)}(x_i) \geq 0 \qquad \Rightarrow (v_i - x_i)^2 = q_i - 2v_i x_i + v_i^2 \geq 0$$
$$B_1^{(2)}(x_i) \geq 0 \qquad \Rightarrow 2(x_i - u_i)(v_i - x_i) = 2(-q_i + (u_i + v_i)x_i - u_i v_i) \geq 0$$
$$B_2^{(2)}(x_i) \geq 0 \qquad \Rightarrow (x_i - u_i)^2 = q_i - 2u_i x_i + u_i^2 \geq 0$$
$$B_0^{(1)}(x_i)B_0^{(1)}(x_j) \geq 0 \Rightarrow (v_i - x_i)(v_j - x_j) = x_{ij} - v_i x_j - v_j x_i + v_i v_j \geq 0$$
$$B_0^{(1)}(x_i)B_1^{(1)}(x_j) \geq 0 \Rightarrow (v_i - x_i)(x_j - u_j) = -x_{ij} + u_j x_i + v_i x_j - v_i u_j \geq 0$$
$$B_1^{(1)}(x_i)B_1^{(1)}(x_j) \geq 0 \Rightarrow (x_i - u_i)(x_j - u_j) = x_{ij} - u_j x_i - u_i x_j + u_i u_j \geq 0.$$

7.2   Inaccuracy Issues

The Bernstein polytope encloses very tightly the underlying algebraic quadratic patch: $(x_1, \ldots, x_n, x_1^2, \ldots, x_n^2, x_1 x_2, \ldots, x_{n-1} x_n)$, $x_i \in [0, 1]$. Thus with a naive floating point implementation, some roots are missed because of rounding errors. For example, when solving $x^2 - x = 0$ with $x \in [0, 1]$, the line $y - x = 0$ is considered, see Fig. 1. If this line becomes $y - x = \epsilon$ with $\epsilon > 0$ due to inaccuracy, the two roots are missed.

For conciseness, we only mention the principle of three solutions: the first and the simplest one is to resort to an exact rational arithmetic; unfortunately it is terribly slow. Michelucci's solver uses this first solution. We then considered a second solution, which resorts to interval arithmetics [1,9]; intervals bounds are floating point numbers, and intervals are rounded outwards at each operation. The used intervals are typically some ULPs large and they only account for the rounding inaccuracy. However, the simplex algorithm has to be modified so it is impossible to use pre-existing LP solvers in floating point arithmetic. For this reason, we prefered a third approach: the solution error of the final linear system in the LP solver is bounded by a backwards error analysis à la Wilkinson, and the LP inequalities are changed accordingly. Fünfzig's solver used this approach [6].

7.3   Guarantees and Theorems

Interval solvers use procedures to prove that the studied box contains no root, or contains at least one root, or contains a unique regular root. These tests rely on mathematical theorems, e.g., Miranda or Kantorovich. This section presents mathematical theorems, which fit well with TBB solvers, including the new solver in Sect. 6. Details will be given elsewhere in a forthcoming article.

Some solvers require an *existence test*. Poincaré-Miranda's theorem, can be used in TBB solvers [11] to prove that a given box contains at least one root of a system of equation. This theorem states under mild assumptions (*i.e.*, the continuity of the functions f_i) that if n continuous functions from $\mathbb{R}^n$ to $\mathbb{R}^n$ are such that each function $f_i(x)$ is always negative on the hyperface $x_i = 0$ of the hypercube $[0, 1]^n$ and $f_i(x)$ is always positive on the opposite hyperface $x_i = 1$ for $i = 1, \ldots, n$, then the system $f_1(x) = \ldots = f_n(x) = 0$ has at least one root in the hypercube $[0, 1]^n$. The hypothesis of Miranda's theorem is more likely to hold if the system is preconditioned. Instead of solving the initial system $f(x) = 0$, a linear combination of the f_i is considered so that its jacobian is approximately the identity matrice at the center of the studied box. This preconditioned system is $g(x) = J(x_c)^{-1}f(x) = 0$.

Some solvers require an *uniqueness test*. Newton-Kantorovich's theorem can be used to prove that a box contains a unique regular root of a system of non-linear equations. This theorem is especially convenient for algebraic quadratic systems, where second derivatives are constant. A second computable condition for uniqueness is given by Kim and Elber [5]: they prove that if the null vector is the only common tangent vector to hypersurfaces $f_i(x) = 0$ then the uniqueness of the root is guaranteed. An equivalent condition is that all enclosing cones of normals of the n hypersurfaces $f_i(x) = 0$ are disjoint. After preconditionning, this condition becomes likely for a small enough box enclosing a unique regular root r. In this case, preconditioning makes hypersurfaces close to orthogonal planes passing through r. A third computable condition that guarantees unique-ness considers the Newton map: $n(x) = x - Mf(x)$, where $f(x) = 0$ is the system to be solved and where M is close to the inverse of the jacobian of f at the center of the studied box. It also considers the norm of its jacobian $n'(x) = I - Mf'(x)$. If for some norm $\|n'(x)\| < 1$ in the studied box then $n(x)$ is guaranteed to be contracting in the studied box, which proves that the root is unique. An upper bound of the maximum and infinite norms can be computed with interval anal-ysis. The approach proposed in this article is also able to compute such upper bounds for matrix norms.

Several methods have been proposed to detect quickly that a studied box contains no root [13,11]. The paper [11] takes also into account inequalities $g_i(x) \leq 0$. Its principle is to search with linear programming a polynomial $h = \sum_j \alpha_j f_j + \sum \beta_i g_i$ with $\alpha_j \in \mathbb{R}$ and $\beta_i \geq 0$ such that $h > 1$ in the studied box with the smallest coefficient in the TBB is 1. If such h exists then the studied box contains no root. It turns out that the new solver in Sect. 6 straightforwardly supersedes this method. The studied box contains no root if the feasible set of the LP problem is empty.

7.4   Solver Implementation

Dominique Michelucci implemented the first variant of the new solver described in Sect. 6 in May 2008. He used exact rational arithmetic to avoid errors due to numerical inaccuracy, and a straightforward simplex solver [2] with ratio-nal arithmetic in Ocaml. This implementation shows the feasibility of the LP reduction approach but the solver is too slow in practice.

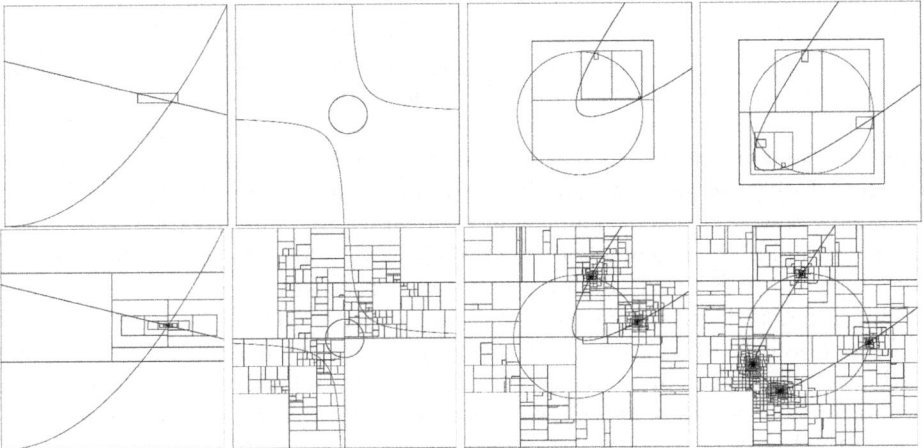

Fig. 5. Comparison of the new solver and a standard interval Newton solver on the same 2D examples.   Top: boxes computed with the new solver.   Bottom: boxes computed with a standard interval Newton solver. Clearly, the new solver earlier detects empty boxes, and its convergence rate is better.

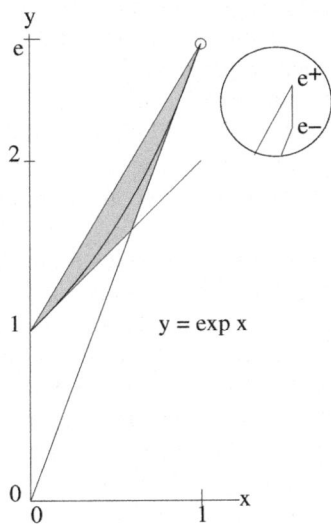

Fig. 6. A convex polygon enclosing the section of the curve $(x, y = \exp x)$, $0 \leq x \leq 1$. e is enclosed in an interval $[e^-, e^+]$.

Then, Christoph Fünfzig implemented the first floating-point variant of this solver in January-June 2009 during his postdoctorate in Dijon [6]. For solving LP problems, Christoph's solver relies on the freely available revised simplex solver SoPlex 1.4.1 developed by Roland Wunderling [18] in his PhD thesis. The solver needs only floating point arithmetic and routinely solves non-linear systems with several dozens of non-linear algebraic (quadratic or higher degrees) equations and unknowns, which previous Bernstein based solvers are not able to solve. In [6], we generate quadratic systems with arbitrary size from circle-packing representations of planar and completely triangulated graphs. [6] gives other examples from geometric constraints in 3D, like the molecule problem, for instance the Stewart platform also called the octahedron problem, or the computation of lines tangent to four given spheres in 3D.

Examples in 2D can be drawn and permit to visually compare the new solver with a standard interval Newton solver. Figure 5 from [6] shows the computation of intersection points between two conics, in the top row with the new solver, and in the bottom row with a standard interval Newton solver.

Both solvers run on the CPU. We are considering the project of GPU implementations: for each box reduction, the $2n$ LP problems can be solved in parallel. Moreover, there is some intrinsic parallelism in the simplex method and in interior-point LP solvers. Thus, a GPU implementation may divide the running time by more than $2n$, where n is the number of unknowns.

8 Conclusion and Future Work

This article has proposed the first polynomial time method to overcome the difficulty due to the exponential cardinality of the TBB. It has defined the Bernstein polytope, which is only implicit in previous TBB solvers. Here are possible future works and concerns, which could not be discussed in detail:

Examples of geometric constraints solving and implementation issues are discussed elsewhere [6] such as accounting for inaccuracy in the simplex method.

GPU implementations of the new solver are possible and planned for a near future.

The Newton-Kantorovich theorem provides a simple and convenient test to prove the uniqueness of a regular root in a box, especially for quadratic systems where all second derivatives are constant. This issue, not specific to TBB solvers, could not be discussed here, as well as deciding about the existence of at least one root in a given box, e.g., the test relying on Miranda's theorem [11].

In [7], we compare the Bernstein polytope with other polytopes like the TBB polytope. The TBB polytope gives exactly the same bounds as the smallest and the largest TBB coefficient. The TBB polytope is tighter than the Bernstein polytope but it has exponential size and complexity: both, its number of hyperplanes and its number of vertices are exponential. [7] concludes that the Bernstein polytope is the best compromise.

The Bernstein polytope should be defined for higher degrees and for other geometric bases, e.g., spline bases. It will extend the scope of the geometric solver by Kim and Elber [5].

The new solver can be generalized in order to manage non-algebraic equations, using for instance transcendental functions cos, exp, etc. It suffices to compute a convex polygone enclosing the 2D curve $(x, \cos x)$ and the 2D curve $(x, \exp x)$ for $x \in [a, b]$. Figure 6 shows a possible convex polygon for $(x, \exp x)$ for $x \in [0, 1]$. This feature is a great advantage compared to other solvers, *e.g.*, homotopy.

The new solver applies without modification to over-constrained systems, where inaccuracy is a serious issue again, but more work is needed to extend it to under-constrained systems, and to compute in a certified manner the topology of semi-algebraic or semi-analytic sets defined by a system of equations and inequalities [3,14]. However, we already use the two solvers to compute tight covers of curves and surfaces [6].

Acknowledgements. We thankfully acknowledge the Regional Council of Burgundy for funding the postdoc position of Ch. Fünfzig at the LE2I in Dijon. This funding has been essential.

References

1. Beaumont, O.: Algorithmique pour les intervalles. Ph.D. thesis, Université de Rennes 1 (1999)
2. Cormen, T.H., Leiserson, C.E., Rivest, R.L., Stein, C.: Introduction to algorithms, 2nd edn. MIT Press, Cambridge (2001)
3. Delanoue, N., Jaulin, L., Cottenceau, B.: Guaranteeing the homotopy type of a set defined by nonlinear inequalities. Reliable Computing 13(5), 381–398 (2007)
4. Durand, C.B.: Symbolic and Numerical Techniques for Constraint Solving. Ph.D. thesis, Purdue University (1998)
5. Elber, G., Kim, M.-S.: Geometric constraint solver using multivariate rational spline functions. In: SMA 2001: Proc. of the 6th ACM Symp. on Solid Modeling and Applications, pp. 1–10. ACM Press, New York (2001), doi:10.1145/376957.376958
6. Fünfzig, C., Michelucci, D., Foufou, S.: Nonlinear systems solver in floating-point arithmetic using lp reduction. In: SPM 2009: 2009 SIAM/ACM Joint Conference on Geometric and Physical Modeling, pp. 123–134. ACM, New York (2009), doi:10.1145/1629255.1629271
7. Fünfzig, C., Michelucci, D., Foufou, S.: Optimizations for bernstein-based solvers using domain reduction. In: CD Proceedings of Eighth International Symposium on Tools and Methods of Competitive Engineering (TMCE 2010). Faculty of Industrial Design Engineering, Delft University of Technology, Ancona, Italy (2010)
8. Garloff, J., Smith, A.P.: Investigation of a subdivision based algorithm for solving systems of polynomial equations. Journal of Nonlinear Analysis: Series A Theory and Methods 47(1), 167–178 (2001)
9. Kearfott, R.B.: Rigorous Global Search: Continuous Problems. Kluwer, Dordrecht (1996)
10. Martin, R., Shou, H., Voiculescu, I., Bowyer, A., Wang, G.: Comparison of interval methods for plotting algebraic curves. Computer Aided Geometric Design 7(19), 553–587 (2002), citeseer.ist.psu.edu/article/martin02comparison.html
11. Michelucci, D., Foufou, S.: Bernstein basis for interval analysis: application to geometric constraints systems solving. In: Bruguera, Daumas (eds.) Proceedings of 8th Conference on Real Numbers and Computers, pp. 37–46. Unidixital, Santiago de Compostela (2008)

12. Michelucci, D.: Solving geometric constraints by homotopy. IEEE Trans on Visualization and Computer Graphics, 28–34 (1996)
13. Mourrain, B., Pavone, J.-P.: Subdivision methods for solving polynomial equations. Journal of Symbolic Computation 3(44), 292–306 (2009)
14. Delanoue, N., Jaulin, L., Cottenceau, B.: Using interval arithmetic to prove that a set is path-connected. Theoretical Computer Science, Special issue: Real Numbers and Computers 351(1), 119–128 (2006)
15. Reuter, M., Mikkelsen, T.S., Sherbrooke, E.C., Maekawa, T., Patrikalakis, N.M.: Solving nonlinear polynomial systems in the barycentric bernstein basis. Vis. Comput. 24(3), 187–200 (2008)
16. Sherbrooke, E.C., Patrikalakis, N.M.: Computation of the solutions of nonlinear polynomial systems. Comput. Aided Geom. Des. 10(5), 379–405 (1993)
17. Sommese, A.J., Wampler, C.W.: Numerical solution of polynomial systems arising in engineering and science. World Scientific Press, Singapore (2005)
18. Wunderling, R.: Paralleler und objektorientierter Simplex-Algorithmus. Ph.D. thesis, TU Berlin (1996), http://www.zib.de/Publications/abstracts/TR-96-09/

Offsetting Revolution Surfaces*

Fernando San Segundo and J. Rafael Sendra

Universidad de Alcalá, Depto. de Matemáticas,
E-28871 Alcalá de Henares, Madrid, Spain
{fernando.sansegundo,rafael.sendra}@uah.es

Abstract. In this paper, first, we provide a resultant-based implicitization method for revolution surfaces, generated by non necessarily rational
curves. Secondly, we analyze the offsetting problem for revolution surfaces, proving that the offsetting and the revolution constructions are
commutative. Finally, as a consequence of this, the (total and partial)
degree formulas for the generic offset to an irreducible plane curve, given
in our previous papers, are extended to the case of offsets to surfaces of
revolution.

Keywords: offset, revolution surface, implicit equation.

1   Introduction

Revolution surfaces are very common objects in Computer Aided Geometric Design, and offsetting a surface is also a frequently used process in the applications.
Thus, it is natural to study the offsetting process for these special surfaces. In
the Geometric Modeling literature, revolution surfaces are often introduced in-
formally, and under the assumption that they are generated by a rational plane
curve (see e.g. [1], [4], [6]). Here we address the more general situation, in which
the generating curve is any algebraic plane curve $\mathcal{C}$, given by its implicit equation.

In order to do this, in the first part of our work, we introduce a formal notion
of surface of revolution by means of incidence diagrams, and from there we state
some preliminary properties. Then we show how the implicit equation of the
revolution surface is related to the implicit equation of the initial curve by means
of resultants. This result shows that, even when the generating curve is a rational
curve given parametrically, an efficient way to obtain the implicit equation of
a revolution surface is to apply the most suitable curve implicitization method,
and then use the result in Theorem 6.

In the second part, we apply the above ideas and results to the offsetting
process in the case of revolution surfaces. The main result of this part is Theorem 11, where we prove that the offset of a revolution surface is the surface of
revolution of the offset curve. From this result, many properties of the offset to
a surface of revolution may be traced back to the properties of the generating

---

* This work has been partially supported by Research Project MTM2008-04699-C03-
01 of the Spanish Ministerio de Ciencia e Innovación.

T. Sturm and C. Zengler (Eds.): ADG 2008, LNAI 6301, pp. 179–188, 2011.

curve. Here we focus on the degree problem for offset surfaces. Thus, we show how the formulae in [7] and [9] generalize to surfaces of revolution. This provides, to our knowledge, the first example in the literature of offset degree analysis for a certain family of surfaces.

Some of these results have been presented, in the form of a short communication, at EACA 2008, held at Granada, Spain, in September 2008, and a short version of this work appears in the book of abstracts of that conference (see [8]).

2   Surfaces of Revolution

In the sequel, $\mathbb{K}$ is an algebraically closed field of characteristic zero. For the application to CAGD, one considers $\mathbb{K} = \mathbb{C}$ as the algebraic closure of $\mathbb{R}$. Let $\mathcal{C}$ be an algebraic irreducible plane affine curve (seen in the coordinate (y_2, y_3)–plane) defined by the irreducible polynomial $f(y_2, y_3) \in \mathbb{K}[y_2, y_3]$, and not equal to the line $y_2 = 0$ (this is because we will rotate around this line). The construction is illustrated in Fig. 1.

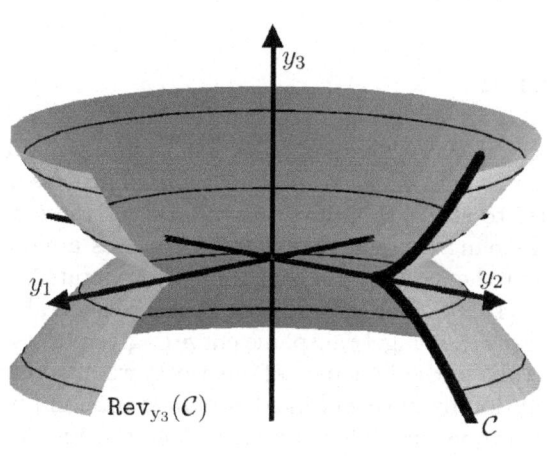

Fig. 1. Construction of a Surface of Revolution

Formally, in order to introduce the revolution construction, we consider the following incidence diagram:

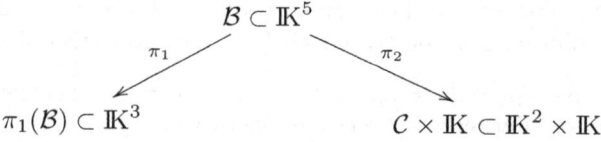

where, denoting $\bar{y} = (y_1, y_2, y_3)$, the revolution incidence variety is

$$\mathcal{B} = \left\{ (r, \bar{y}, \lambda) \in \mathbb{K}^5 \left| \begin{array}{l} f(r, y_3) = 0, \\ r^2 = y_1^2 + y_2^2, \\ (1 + \lambda^2)y_1 = 2\lambda r, \\ (1 + \lambda^2)y_2 = (1 - \lambda^2)r. \end{array} \right. \right\},$$

with this projection maps:

$$\pi_1 : \qquad \mathbb{K}^5 \quad \longrightarrow \quad \mathbb{K}^3, \qquad\qquad \pi_2 : \qquad \mathbb{K}^5 \quad \longrightarrow \quad \mathbb{K}^2 \times \mathbb{K}$$
$$(r, y_1, y_2, y_3, \lambda) \longmapsto (y_1, y_2, y_3) \qquad\qquad (r, y_1, y_2, y_3, \lambda) \longmapsto ((r, y_3), \lambda).$$

With this notation we are ready for the formal definition:

Definition 1. *The surface of revolution generated by rotating $\mathcal{C}$ around the y_3 axis is the Zariski closure $\pi_1(\mathcal{B})^*$ of $\pi_1(\mathcal{B})$. We denote the surface of revolution of $\mathcal{C}$ by* $\mathrm{Rev}_{y_3}(\mathcal{C})$.

The following lemma lists the properties of the incidence diagram that we will need in the sequel:

Lemma 2. *Let $\mathcal{C}$ be irreducible, and not equal to the line $y_2 = 0$.*

(1) π_2 is a birational map.
(2) π_1 is a finite map.

Proof. First note that, for $((r, y_3), \lambda) \in \mathcal{C} \times (\mathbb{K} \setminus \{\pm\sqrt{-1}\})$, the inverse of π_2 is given by

$$\pi_2^{-1}((r, y_3), \lambda) = \left(r, \frac{2r\lambda}{\lambda^2 + 1}, \frac{(1 - \lambda^2)r}{\lambda^2 + 1}, y_3, \lambda \right).$$

On the other hand, since $\mathcal{C}$ is not the line $y_2 = 0$, we can find a point (in fact infinitely many) $(r_0, c) \in \mathcal{C}$ with $r_0 \neq 0$. Then the point $p = (r_0, 0, r_0, c, 0)$ is in $\mathcal{B}$, and since $\pi_1(r_0, 0, r_0, c, 0) = (0, r_0, c)$, the fiber $\pi_1^{-1}(\pi_1(p))$ is determined by the system:

$$\left\{ f(r, c) = 0, \; r^2 = 0^2 + r_0^2, \; (1 + \lambda^2)0 = 2\lambda r, \; (1 + \lambda^2)r_0 = (1 - \lambda^2)r \right\}.$$

Therefore, because of $r_0 \neq 0$, we must have $\lambda = 0$ and $r = r_0$. The fiber is thus zero-dimensional (see Theorem 11.12 in [5]). □

Remark 3. Note that if $\mathcal{C}$ is the line $y_2 = 0$, then *all the points* in $\mathcal{B}$ are of the form $p = (0, a, b, c, \lambda_0)$ with $a^2 + b^2 = 0$ or $\lambda_0^2 + 1 = 0$. In the first case, $\pi_1^{-1}(\pi_1(p))$ contains all the points $(0, a, b, c, \mu)$ for any $\mu \in \mathbb{K}$. In the second case, $\pi_1^{-1}(\pi_1(p))$ contains all the points $(0, \alpha, \beta, c, \lambda_0)$ with $\alpha^2 + \beta^2 = 0$. In any case the fiber is 1-dimensional, and it follows that $\dim(\mathrm{Rev}_{y_3}(\mathcal{C})) = \dim(\mathcal{B}) - 1 = 1$.

The following proposition, which is a direct consequence of the previous lemma, shows that the above notion of revolution surface is well defined.

Proposition 4. *Let $\mathcal{C}$ be irreducible, and not equal to the line $y_2 = 0$. Then $\mathrm{Rev}_{y_3}(\mathcal{C})$ is an irreducible surface.*

Remark 5. Note that, if $\mathcal{C}$ is not irreducible, then its surface of revolution can be introduced as the union of the surfaces of revolution of its components.

3   Implicitization of Revolution Surfaces

Our next goal is to derive a method for computing the implicit equation of $\mathrm{Rev}_{y_3}(\mathcal{C})$. For this purpose, first, collecting terms of odd and even degree in y_2, we write f (i.e. the implicit equation of $\mathcal{C}$) as follows:

$$f(y_2, y_3) = A(y_2^2, y_3) + y_2 B(y_2^2, y_3).\tag{1}$$

We will see that there are two cases to consider:

- case (a): either $B = 0$ and hence $f \in \mathbb{K}[y_2^2, y_3]$; that is, f contains only even powers of y_2,
- case (b): or $B \neq 0$, when f contains at least one odd power of y_2.

Then, the following theorem shows how the implicit equations of $\mathrm{Rev}_{y_3}(\mathcal{C})$ and $\mathcal{C}$ are related by means of resultants.

Theorem 6. *Let $\sigma(y_1, y_2, y_3)$ be the implicit equation of $\mathrm{Rev}_{y_3}(\mathcal{C})$, and let the implicit equation of $\mathcal{C}$ be $f(y_2, y_3)$. Then there exists $\ell \in \mathbb{N}$ such that*

$$\sigma(y_1, y_2, y_3)^\ell = \mathrm{Res}_r(f(r, y_3), r^2 - (y_1^2 + y_2^2)).$$

Furthermore, if $R(y_1, y_2, y_3)$ is the above resultant, it holds that:

1. In case (a)*, $R(y_1, y_2, y_3) = (A(y_1^2 + y_2^2, y_3))^2$, and so*

$$\sigma(y_1, y_2, y_3) = A(y_1^2 + y_2^2, y_3).$$

2. In case (b)*,*

$$\sigma(y_1, y_2, y_3) = R(y_1, y_2, y_3) = A^2(y_1^2 + y_2^2, y_3) - (y_1^2 + y_2^2)B^2(y_1^2 + y_2^2, y_3).$$

Proof. Let I be the ideal in $\mathbb{K}[r, y_1, y_2, y_3, \lambda]$ generated as follows:

$$I = \langle f(r, y_3), r^2 - y_1^2 - y_2^2, (1 + \lambda^2)y_1 - 2\lambda r, (1 + \lambda^2)y_3 - (1 - \lambda^2)r \rangle$$

(that is, the ideal generated by the polynomials in $\mathcal{B}$). Observe that the resultant R belongs to the (r, λ)-elimination ideal $I \cap \mathbb{K}[y_1, y_2, y_3]$. Moreover, R equals the product of $f(r, y_3)$ evaluated at the roots of $r^2 - y_1^2 - y_2^2$ as a polynomial in r (see e.g. [12]). So $R = f(\sqrt{y_1^2 + y_2^2}, y_3)f(-\sqrt{y_1^2 + y_2^2}, y_3)$. Therefore, in case (a), $R(\bar{y}) = (A(y_1^2 + y_2^2, y_3))^2$. Furthermore, in case (b),

$$R(\bar{y}) = A^2(y_1^2 + y_2^2, y_3) - (y_1^2 + y_2^2)B^2(y_1^2 + y_2^2, y_3).$$

Now, in either case, since σ is irreducible, it divides R. And conversely, using that $r^2 - y_1^2 - y_2^2$ is monic in r, by the Extension Theorem for resultants (see [3]) one deduces that $R(\bar{y})$ defines a surface contained in $\mathrm{Rev}_{y_3}(\mathcal{C})$. Thus, σ is the square-free part of R. $\qquad\square$

Remark 7. If $\mathcal{C}$ is not irreducible, the method described in this theorem still provides the implicit equation of its surface of revolution; in this case, the square-free part of the resultant factors into the implicit equations of the components of $\mathrm{Rev}_{y_3}(\mathcal{C})$.

Next theorem, that is a direct consequence of the previous theorem, gives a complete degree analysis of $\mathrm{Rev}_{y_3}(\mathcal{C})$.

Theorem 8. *Let $\sigma(y_1, y_2, y_3)$ be the implicit equation of $\mathrm{Rev}_{y_3}(\mathcal{C})$, and let the implicit equation of $\mathcal{C}$ be $f(y_2, y_3)$. Then, it holds that:*

1. *In case (a), $\deg_{(y_1, y_2, y_3)}(\sigma) = \deg_{(y_2, y_3)}(f)$, $\deg_{y_i}(\sigma) = \deg_{y_2}(f)$, for $i = 1, 2$, and $\deg_{y_3}(\sigma) = \deg_{y_3}(f)$.*

2. *In case (b), $\deg_{(y_1, y_2, y_3)}(\sigma) = 2 \deg_{(y_2, y_3)}(f)$, $\deg_{y_i}(\sigma) = 2 \deg_{y_2}(f)$, for $i = 1, 2$, and $\deg_{y_3}(\sigma) = 2 \deg_{y_3}(f)$.*

We finish this section with an illustrating example.

Example 9. Let $\mathcal{C}$ be the non-rational cubic defined by $f(y_2, y_3) = y_3^2 - y_2(y_2^2 - 1)$. Then $A(y_2, y_3) = y_3^2$ and $B(y_2, y_3) = -(y_2^2 - 1) \neq 0$, and so $f(y_2, y_3)$ is in case (b). Thus, the implicit equation of $\mathrm{Rev}_{y_3}(\mathcal{C})$ is given by:

$$\sigma(y_1, y_2, y_3) = y_3^4 - (y_1^2 + y_2^2)((y_1^2 + y_2^2) - 1)^2.$$

4   Offsets to Revolution Surfaces

In this section we apply the above results to analyze the offsetting process in the case of revolution surfaces. Let $\mathcal{C}$ be a curve as above. We denote by $\mathcal{O}_d(\mathcal{C})$ the offset to $\mathcal{C}$ at distance d (see [2]). The normal vectors to $\mathrm{Rev}_{y_3}(\mathcal{C})$ have the following (geometrically intuitive) fundamental property.

Lemma 10. *Let $\tilde{p} \in \mathrm{Rev}_{y_3}(\mathcal{C})$ be obtained rotating $p \in \mathcal{C}$ around the y_3 axis, and let us denote by θ the particular rotation carrying p to $\tilde{p}$. Then, $\tilde{N}(\tilde{p})$, the normal vector to $\mathrm{Rev}_{y_3}(\mathcal{C})$ at $\tilde{p}$, is parallel to the vector $\theta(\tilde{N}(p))$, obtained by applying the same rotation to the normal vector $N(p)$ to $\mathcal{C}$ at p.*

Proof. With the notation introduced for the revolution incidence variety, note that if $p = (0, r, y_3) \in \mathcal{C}$ and $(r, y_3, \lambda) \in \mathcal{B}$, then the rotation in the statement is given by the following matrix:

$$M = \begin{pmatrix} \dfrac{1 - \lambda^2}{\lambda^2 + 1} & \dfrac{2\lambda}{\lambda^2 + 1} & 0 \\[2mm] \dfrac{-2\lambda}{\lambda^2 + 1} & \dfrac{1 - \lambda^2}{\lambda^2 + 1} & 0 \\[2mm] 0 & 0 & 1 \end{pmatrix},$$

and therefore $\tilde{p} = \theta(p) = \left(\frac{2r\lambda}{\lambda^2+1}, \frac{(1-\lambda^2)r}{\lambda^2+1}, y_3\right)$. Then, from (1), it is easy to check that a normal vector to $\mathcal{C}$ at p is given by:

$$\tilde{N}(p) = [0, 2\,\partial_1 A(\nu)r + B(\nu) + 2\,r^2\partial_1 B(\nu), \partial_2 A\left(r^2, y_3\right) + r\partial_2 B(\nu)],$$

where $\nu = \left(r^2, y_3\right)$ and $\partial_1 A, \partial_2 A$ denote the partial derivatives of A w.r.t. its first and second variable, respectively (similarly for B). The rotation of this normal vector, $\theta(\tilde{N}(p))$, is given by $M \cdot N(p)$, where M is the above matrix. On the other hand, computing the gradient $\nabla\sigma(\tilde{p})$, and taking into account that $\sigma(\tilde{p}) = 0$, a straightforward computation shows that:

$$\nabla\sigma(\tilde{p}) \wedge M \cdot N(p) = 0,$$

where $\wedge$ denotes cross product. This concludes the proof of our claim.   □

In the following theorem, which is a direct consequence of the above reasonings, we assume that both $\mathcal{C}$ and $\mathcal{O}_d(\mathcal{C})$ are in the (y_2, y_3)–plane.

Theorem 11. $\mathcal{O}_d(\mathrm{Rev}_{y_3}(\mathcal{C})) = \mathrm{Rev}_{y_3}(\mathcal{O}_d(\mathcal{C}))$.

Now we turn to the degree problem. From the last theorem, if we can deduce in which case ((a) or (b)) of Theorem 6) the implicit equation of $\mathcal{O}_d(\mathcal{C})$ is, then applying Theorem 8 as well as results in [7] and [9], we can provide formulae for the partial and total degree of $\mathcal{O}_d(\mathrm{Rev}_{y_3}(\mathcal{C}))$. Note that polynomials in $\mathbb{K}[y_2^2, y_3]$ (that is, the polynomials in case (b)) are characterized by the symmetry condition $f(-y_2, y_3) = f(y_2, y_3)$. Thus, we need to show that this symmetry condition is inherited by the offset. The answer is contained in the following two propositions. The first one analyzes the problem from the implicit point of view. The second one shows how to detect this property from the parametric point of view. This is useful e.g. if one is given a parametric representation of the generating curve, and wishes to obtain the offset surface degrees without implicitizing the curve.

In the following proposition, let $f(y_2, y_3)$ be the polynomial defining $\mathcal{C}$ and let $g(y_2, y_3, d)$ be the generic equation of the offset $\mathcal{O}_d(\mathcal{C})$ (see [9] for its definition and properties).

Proposition 12. Let $f(y_2, y_3)$ and $g(y_2, y_3, d)$ be the polynomials defining $\mathcal{C}$ and $\mathcal{O}_d(\mathcal{C})$, respectively. Then, $g(y_2, y_3, d) = g(-y_2, y_3, d)$ if and only if $f(y_2, y_3) = f(-y_2, y_3)$.

Proof. The right-left implication follows from the offset geometric construction, because the normal vector to $\mathcal{C}$ and the normal vector to its offset at the corresponding points are parallel. Conversely, suppose that $g(y_2, y_3, d) = g(-y_2, y_3, d)$. Now, let d_0 be such that no coefficient w.r.t. $\{y_2, y_3\}$ of g vanishes when substituting d by d_0, and such that $g(y_2, y_3, d_0)$ is the implicit equation of $\mathcal{O}_{d_0}(\mathcal{C})$. Then $\mathcal{O}_{d_0}(\mathcal{O}_{d_0}(\mathcal{C})) = \mathcal{C} \cup \mathcal{O}_{2d_0}(\mathcal{C})$ (see [10]). Let $\tilde{g}(y_2, y_3, d_0)$ be the polynomial defining $\mathcal{O}_{d_0}(\mathcal{O}_{d_0}(\mathcal{C}))$. Then $\tilde{g}(y_2, y_3, d_0) = g(y_2, y_3, 2d_0)f(y_2, y_3)$. Furthermore, because of the hypothesis and how d_0 has been taken, $g(y_2, y_3, 2d_0) = g(-y_2, y_3, 2d_0)$.

Moreover, because of the first implication $\tilde{g}(y_2, y_3, d_0)$ inherits this property. Now, from $\tilde{g} = gf$, it follows immediately that $f(y_2, y_3) = f(-y_2, y_3)$.   □

Now we show how to detect this symmetry from a parametrization of $\mathcal{C}$. So we assume that $\mathcal{C}$ is rational, and that

$$P_{\mathcal{C}}(t) = \left(\frac{C_1(t)}{C_3(t)}, \frac{C_2(t)}{C_3(t)} \right), \text{ with } \gcd(C_1, C_2, C_3) = 1, \tag{2}$$

is a rational parametrization of $\mathcal{C}$. Then, the symmetry condition can be translated into asking that

$$\tilde{P}_{\mathcal{C}}(s) = \left(-\frac{C_1(s)}{C_3(s)}, \frac{C_2(s)}{C_3(s)} \right)$$

also parametrizes $\mathcal{C}$. So, taking into account that the implicit equation of $\mathcal{C}$ is the square-free part of $\mathrm{Res}_t(C_3 y_2 - C_1, C_3 y_3 - C_2)$ (see [11]), one gets the following result.

Proposition 13. *Let $g(y_2, y_3, d)$ be the generic offset equation for $\mathcal{O}_d(\mathcal{C})$. Then $g(y_2, y_3, d) = g(-y_2, y_3, d)$ iff*

$$\gcd(C_3(t)C_1(s) + C_1(t)C_3(s), C_3(t)C_2(s) - C_2(t)C_3(s))$$

is non-trivial.

Proof. Let

$$M_1(s, t) = C_3(t)C_1(s) + C_3(s)C_1(t), \quad M_2(s, t) = C_3(t)C_2(s) - C_3(s)C_2(t),$$

and $D(s, t) = \gcd(M_1, M_2)$. We first observe that M_1 and M_2 can not be both simultaneously zero, since this would imply that $P_{\mathcal{C}}$ is not a parametrization. Moreover, note that if either $\frac{C_1}{C_3}$ or $\frac{C_2}{C_3}$ is constant, the result follows. Thus, in the rest of the proof we assume that no component of P is constant.

Let $\mathcal{D} \subset \mathbb{K}^3$ be defined as follows:

$$\mathcal{D} := \{(t_o, s_o, u_o) \in \mathbb{K}^3 \mid u_o C_3(t_o)C_3(s_o) = 1, M_1(t_o, s_o) = M_2(t_o, s_o) = 0\}.$$

We consider the diagram:

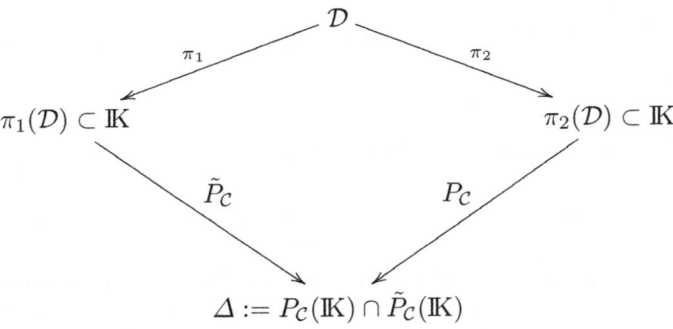

where $\pi_1(t_o, s_o, u_o) = t_o, \pi_2(t_o, s_o, u_o) = s_o$ and $\tilde{P}_C(s) = \left(-\frac{C_1(s)}{C_3(s)}, \frac{C_2(s)}{C_3(s)}\right)$ is as in (2). We observe that the diagram is commutative and that $P_C \circ \pi_1$ and $\tilde{P}_C \circ \pi_2$ are both surjective on Δ. Let D be constant. Then $\mathcal{D}$ is either empty, or zero-dimensional. Thus, Δ is either empty or zero-dimensional. In particular, $\tilde{P}_C$ does not parametrize $\mathcal{C}$, and by Proposition 12 we conclude that $f(y_2, y_3) \neq f(-y_2, y_3)$. If D is a non-constant polynomial, we first observe that

$$\gcd(D, C_3(t)) = \gcd(D, C_3(s)) = 1.$$

Indeed, if $\gcd(D, C_3(t)) \neq 1$ (similarly if $\gcd(D, C_3(s)) \neq 1$), then $M_1(t, s)$ and $M_2(t, s)$ have a non-trivial common factor depending only on t. Taking into account that no component of P is constant, that factor would then divide $C_1(t), C_2(t)$ and $C_3(t)$, which is impossible because $\gcd(C_1, C_2, C_3) = 1$ by hypothesis. In this situation we have that $\dim(\mathcal{D}) = 1$, and that $\mathbb{K} \setminus \pi_1(\mathcal{D})$ is empty or finite. The same holds for $\mathbb{K} \setminus \pi_2(\mathcal{D})$. Thus, $\dim(\Delta) = 1$. This implies, by Proposition 12, that $f(y_2, y_3) = f(-y_2, y_3)$. □

Using these results, one derives the following algorithm for the solution of the offset degree problem in the case of surfaces of revolution.

Algorithm:
Offset Degree for the Surface of Revolution Generated by the Curve $\mathcal{C}$.

- <u>Input</u>: Either the defining polynomial f or a rational parametrization $P_C(t)$, as above, of $\mathcal{C}$ ($\mathcal{C}$ is not the axes $y_2 = 0$).
- <u>Output</u>: The total and partial degrees of $\mathcal{O}_d(\text{Rev}_{y_3}(\mathcal{C}))$.

1. Apply either Proposition 12 or 13 to check whether the generic offset equation $g(y_2, y_3, d)$ of $\mathcal{O}_d(\mathcal{C})$ is in case(a) or case(b) (with the terminology introduced before Theorem 6).
2. Apply formulae in [7], [9] to get

$$\delta = \deg_{y_2, y_3}(g), \delta_2 = \deg_{y_2}(g), \delta_3 = \deg_{y_3}(g), \delta_d = \deg_d(g).$$

3. Let $G(y_1, y_2, y_3, d)$ be the polynomial defining $\mathcal{O}_d(\text{Rev}_{y_3}(\mathcal{C}))$.
   In case (a) return:
   $\deg_{\bar{y}}(G) = \delta$, $\deg_{y_1}(G) = \deg_{y_2}(G) = \delta_2$, $\deg_{y_3}(G) = \delta_3$, $\deg_d(G) = \delta_d$.
   In case (b), return:
   $\deg_{\bar{y}}(G) = 2\delta$, $\deg_{y_1}(G) = \deg_{y_2}(G) = 2\delta_2$, $\deg_{y_3}(G) = 2\delta_3$, $\deg_d(G) = 2\delta_d$.

Let us finish with some examples:

Example 14. Consider the parabola with equation given by $f(y_2, y_3) = y_3 - y_2^2$. Then obviously $f(-y_2, y_3) = f(y_2, y_3)$, and so we are in case (a). Applying the formulae in [7] and [9], one has: $\{\delta = 6, \delta_2 = 6, \delta_3 = 4, \delta_d = 6\}$, and so the algorithm returns $\{\deg_{(y_1, y_2, y_3)}(G) = 6, \deg_{y_1}(G) = \deg_{y_2}(G) = 6, \deg_{y_3}(G) = 4, \deg_d(G) = 6\}$ for the degrees of the offset of the circular paraboloid defined by $y_3 - (y_1^2 + y_2^2) = 0$, which is the surface of revolution generated by $\mathcal{C}$.

Example 15. For the non-rational cubic $\mathcal{C}$ in Example 9, the formulae in [7] and [9], give: $\{\delta = 14, \delta_2 = 14, \delta_3 = 12, \delta_d = 14\}$, and so, since we are in case (b), the algorithm returns

$$\deg_{(y_1,y_2,y_3)}(G) = 28, \deg_{y_1}(G) = \deg_{y_2}(G) = 28, \deg_{y_3}(G) = 24, \deg_d(G) = 28,$$

for the degrees of $\mathrm{Rev}_{y_3}(\mathcal{C})$.

Example 16. Let $\mathcal{C}$ be the lemniscate parametrized by

$$P_{\mathcal{C}}(t) = \left(\frac{\sqrt{2}(t + t^3)}{1 + t^4}, \frac{\sqrt{2}(t - t^3)}{1 + t^4} \right).$$

Then one has

$$\gcd(C_3(t)C_1(s) + C_1(t)C_3(s), C_3(t)C_2(s) - C_2(t)C_3(s)) = ts + 1,$$

and it follows that we are in case (a). Using the formulae in [7] and [9], one has: $\{\delta = \delta_2 = \delta_3 = \delta_d = 12\}$, and so the algorithm returns

$$\{\deg_{(y_1,y_2,y_3)}(G) = \deg_{y_1}(G) = \deg_{y_2}(G) = \deg_{y_3}(G) = \deg_d(G) = 12\}$$

for the degrees of $\mathrm{Rev}_{y_3}(\mathcal{C})$.

Example 17. Let $\mathcal{C}$ be the Folium parametrized by

$$P_{\mathcal{C}}(t) = \left(\frac{3t}{1 + t^3}, \frac{3t^2}{1 + t^3} \right).$$

Then one has

$$\gcd(C_3(t)C_1(s) + C_1(t)C_3(s), C_3(t)C_2(s) - C_2(t)C_3(s)) = 3$$

and we are in case (b). Using the formulae in [7] and [9], one has: $\{\delta = \delta_2 = \delta_3 = \delta_d = 14\}$, and so the algorithm returns

$$\{\deg_{(y_1,y_2,y_3)}(G) = \deg_{y_1}(G) = \deg_{y_2}(G) = \deg_{y_3}(G) = \deg_d(G) = 28\}$$

for the degrees of $\mathrm{Rev}_{y_3}(\mathcal{C})$.

References

1. Agoston, M.K.: Computer Graphics and Geometric Modeling: Implementation and Algorithms. Springer, Heidelberg (2005)
2. Arrondo, E., Sendra, J., Sendra, J.R.: Parametric generalized offsets to hypersurfaces. Journal of Symbolic Computation 23(2-3), 267–285 (1997)
3. Cox, D.A., Little, J.B., O'Shea, D.: Ideals, Varieties, and Algorithms: An Introduction to Computational Algebraic Geometry and Commutative Algebra, 2nd edn. Springer, Heidelberg (1997)

4. Farin, G.E.: Curves and Surfaces for CAGD: A Practical Guide. Morgan Kaufmann, San Francisco (2001)
5. Harris, J.: Algebraic Geometry: A First Course. Springer, Heidelberg (1992)
6. Patrikalakis, N.M., Maekawa, T.: Shape Interrogation for Computer Aided Design and Manufacturing. Springer, Heidelberg (2002)
7. San Segundo, F., Sendra, J.R.: Degree formulae for offset curves. J. Pure Appl. Algebra 195(3), 301–335 (2005), doi:10.1016/j.jpaa.2004.08.026
8. San Segundo, F., Sendra, J.R.: The offset degree problem for surfaces of revolution. In: Proceedings of the XIth Encuentro de Algebra Computacional y Aplicaciones(EACA 2008), Universidad de Granada, pp. 65–68 (2008)
9. San Segundo, F., Sendra, J.R.: Partial degree formulae for plane offset curves. Journal of Symbolic Computation 44(6), 635–654 (2009), doi:10.1016/j.jsc.2008.10.002
10. Sendra, J.R., Sendra, J.: Algebraic analysis of offsets to hypersurfaces. Mathematische Zeitschrift 234(4), 697–719 (2000)
11. Sendra, J.R., Winkler, F., Pérez-Díaz, S.: Rational Algebraic Curves—A Computer Algebra Approach. Springer, Heidelberg (2007)
12. van der Waerden, B.L.: Algebra. Springer, Heidelberg (2003)

An Introduction to Java Geometry Expert*
(Extended Abstract)

Zheng Ye[1], Shang-Ching Chou[2], and Xiao-Shan Gao[3]

[1] Zhejiang GongShang University, Zhejiang, China
[2] Wichita State Univesity, Kansas, U.S.A
[3] KLMM, Institute of Systems Science, Chinese Academy of Sciences, Beijing, China

Abstract. This paper gives a brief introduction to the system Java Geometry Expert (JGEX). This system consists of three parts: the drawing part, the proving and reasoning part, and the most distinctive part – the part for generating *visually dynamic presentation of proofs* in plane geometry. The current version of JGEX is beta 0.80, which is available at our website woody: http://woody.cs.wichita.edu.

Keywords: Plane Geometry, Geometry Theorem Proving, Visually Dynamic Presentation, Pythagoras' Theorem.

1 Introduction

Highly successful algebraic methods for automated geometry theorem proving have been developed since Wu's pioneering work in 1978 [16]. Hundreds of difficult geometry theorems have been proved with these methods [7,9,10,4,15,14].

The proofs, generated by these algebraic methods, involve computations of polynomials with hundreds or even thousands of terms. Thus they are generally not (human) readable.

Most proofs in geometry textbooks are synthetic (possibly with some very simple algebraic computations). Students can read the proofs step by step with assistance of one or more diagrams. However, they often need to spend time and energy on identifying a geometry element in the proof text with that in the corresponding diagram. When the same element is mentioned later in the proof text they might spend equal amounts of time and energy on identifying it again in the diagram. When the diagram becomes complicated, e.g., there are over a dozen of points involved in the diagram, the problem becomes serious not only to novices, but also to experts.

Geometry textbooks generally alleviate this problem by using two or more diagrams with different marks for angles and segments, and possibly with shadowed areas, e.g., a shadowed triangle, in the diagrams. However, this kind of presentation of proofs is static.

With dynamic mediums such as computer displays, we propose an entirely new approach – visually dynamic presentation of proofs (VDPP), to solve this problem.

---

* The work reported here was supported by NSF Grant CCR-0201253.

T. Sturm and C. Zengler (Eds.): ADG 2008, LNAI 6301, pp. 189–195, 2011.

In a single diagram for the proof, when the proof text goes on step by step with mouse clicks, the related geometry elements in the diagram are animated, added, or deleted dynamically with various visually dynamic effects.

We have implemented two methods for generating such presentations of proofs in our developing system Java Geometry Expert (JGEX): the manual input method and the automated method, which will be illustrated by examples in later sections.

2   The Parts of JGEX

The Drawing Part [5]. This part has the dynamic geometry features similar to those used in the popular and excellent systems such as the Geometers Sketchpad, Cabri, and Cinderella.

With mouse clicks, the diagram is constructed and the corresponding geometry statement is generated with its non-degenerate conditions in its *geometric form*. It can be saved in several forms. One form is in plain text, which can be used in turn to generate the diagram with prompting the user to select points.

The Proving and Reasoning Part. Beside the traditional algebraic methods such as Wu's method and the Gröbner basis method, we have also implemented the full-angle method and the deductive database method [3,6] for generating short, elegant synthetic proofs. These are the basis of the next part.

The Part of the Visually Dynamic Representation of Proofs. We have implemented two methods for generation of such visually dynamic representation (VDPP) which will be discussed in detail in next two sections.

3   The Manual Input Method

JGEX provides a very *general* tool for manually creating VDPPs. It can be used by students or teachers to write or to present proofs. We plan to implement four modes. So far we have only implemented Modes 1 and 2.

Mode 1: Animated Diagrams Only. The approach in this mode is very similar to the approach of Proof Without Words (PWW) presented in three excellent books [11,12,13]. However, we add another dimension to the PWW approach, i.e., instead of a static diagram or a series of static diagrams, the diagram here is visually dynamic.

Example 1. A Proof of the Pythagorean Theorem (Fig. 1).

This proof is from the webpage [1] which is a collection of 72 proofs of the Pythagoras theorem. From our visually dynamic presentation in Fig. 1, one can clearly see the elegance of the proof. For the real animated gif file created with JGEX see the Collection at http://woody.cs.wichita.edu/collection, where many JGEX-manually created examples are given, in particular there are over two dozens of proofs of the Pythagorean Theorem with the area dissection

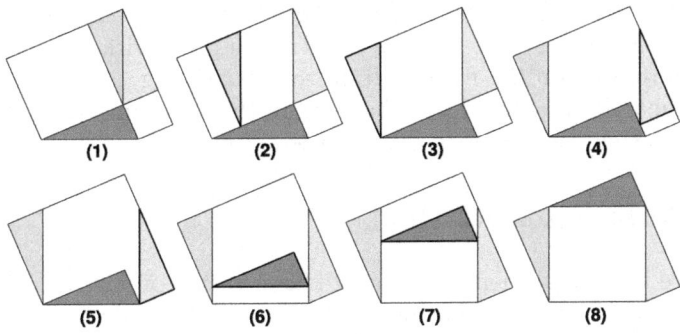

Fig. 1. A Proof of Pythagorean Theorem with Only Three Moves

method, similar to this one, are given. These examples can be easily created with JGEX as a general tool.

Mode 2: Proofs with text and an animation diagram. First a student needs to know the proof. Then he/she inputs the proof step by step mostly with mouse clicks on the diagram to avoid typos (e.g., typing letter A instead of letter S). When the proof is completed, others can see the proof step by step while corresponding geometry elements in the step are animated to reflect the geometric meaning of this step.

In this mode, JGEX only verifies numerical correctness of the assertion of a step by randomly generating many floating point number instances of the diagram. It does not care whether an assertion is a logical consequence of previous assertions. It could be a good tool for students to write proofs. However, the teachers need to decide whether the proof is correct or complete.

Example 2. Let circle O be the circumscribed circle of an equilateral triangle ABC and E a point on the arc AB. Prove that $EC = EA + EB$ (Fig. 2).

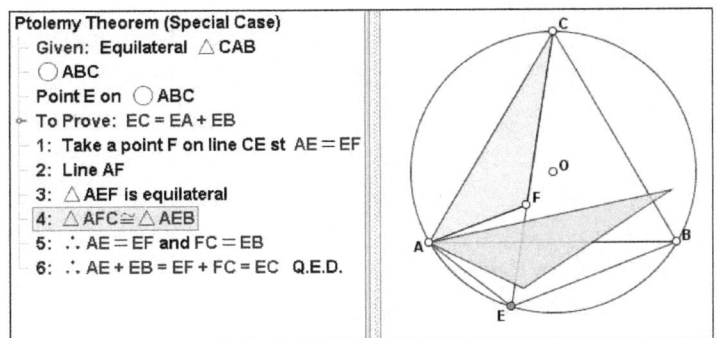

Fig. 2. A Proof of the Special Case of Ptolemy's Theorem

In Fig. 2, the fact $\triangle AFC \cong \triangle AEB$ is highlighted with the visual effects as follows: triangle $\triangle AFC$ and triangle $\triangle AEB$ are filled with colors and a copy of $\triangle AFC$ is rotated on the fly and drops to $\triangle AEB$. Fig. 2 is a static diagram from the animated diagram of this theorem. In this diagram, the red color-filled triangle is rotating and about to drop to the $\triangle AEB$.

4   The Automated Methods

In JGEX, we have implemented automated generation of VDPPs with the full-angle method [3] and with the deductive database method [6]; for details, see our paper [17] which is already published. The two methods developed in the 1990s imply automated addition of auxiliary geometric elements: given two points A and B, three non-collinear points C, D and E, or two lines l_1 and l_2, if the automatically generated proof requires, there will be a line AB or a segment AB, or a circle CDE, or a full-angle $\angle[l_1, l_2]$, etc. However, the methods are unable to add a point of intersection of, say, two given lines.

The full-angle method is a natural way to generate proofs with hierarchical structures. Any non-initial facts found by forward chaining can be expanded to view the proof of the fact for further investigation. If the proof of this fact has a sufficient number of steps, we can consider this fact and its proof as a lemma application. Hierarchically structured proofs allow the user to concentrate on the main steps.

Example 3. (Simson's Theorem) Let E be a point on the circumscribed circle(D) of triangle ABC. Let F, G, and H be the feet of the three perpendicular lines

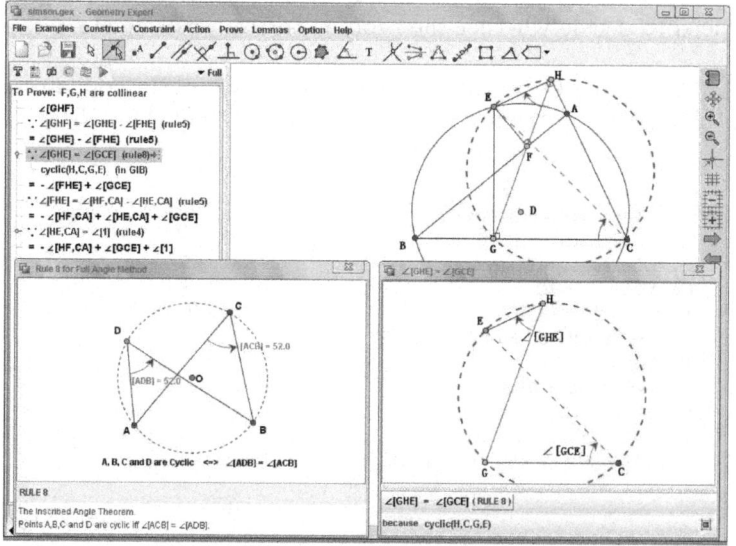

Fig. 3. Simson's Theorem

from point E to the three sides AB, BC, and AC, respectively. Show that F, G and H are collinear (Fig. 3).

Fig. 3 shows the machine-generated proofs with the full-angle method. Step 2 is expanded and highlighted with the two auxiliary angles and one auxiliary circle appear and blink. This step uses Rule 8 (See Fig. 3) with the fact $cyclic(C, E, G, H)$. This fact is found by the forward chaining, i.e., a fact in the fixpoint. The user can expand this step to view the proof of this fact.

There are two floating windows in Fig. 3. The right one shows the portion of the diagram where the rule applies. The left one gives the detail of the rule. In this case, it shows that Rule 8 is the full-angle version of the inscribed angle theorem.

5   Visualization of Fixpoints

The fixpoint generated by the deductive database method contains surprisingly rich amounts of information, some of which is very unexpected. Visualizing fixpoints can help users to explore properties that they are not aware of.

Example 4. (The Orthocenter Theorem) Let CD and BE be two altitudes of triangle ABC, F the intersection of CD and BE, and G the intersection of AF and BC. Show that $\angle[DGA] = \angle[AGE]$ (Fig. 4).

Fig. 4 shows the fixpoint of this theorem found by forward chaining. There are seven groups of angle congruence. By clicking one of them (highlighted in

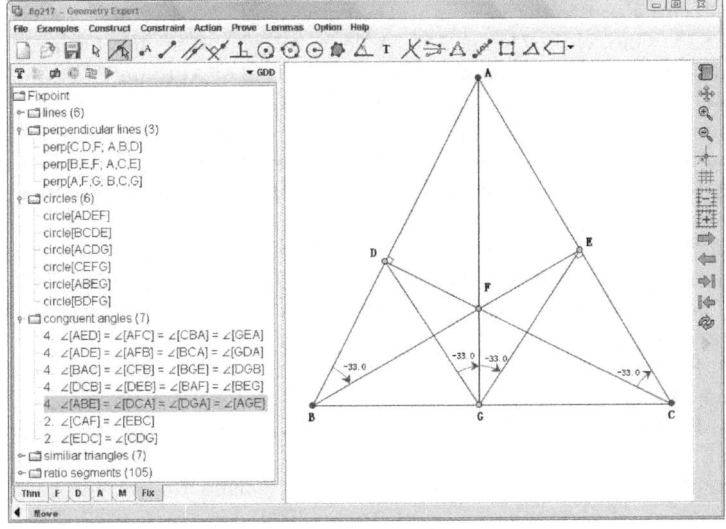

Fig. 4. The Orthocenter Theorem

Fig. 4), the corresponding angles appear in the diagram. We can see that the fact ($\angle[DGA] = \angle[AGE]$) is in the fixpoint thus the conclusion is reached by forward chaining.

6   Conclusion

JGEX is based on our previous version of Geometry Expert (GEX)[8]. However, it has been rewritten completely in Java with emphasis on its ease of use. The most distinctive feature of JGEX is its visually dynamic presentation of proofs. This makes JGEX a valuable tool for generating and presenting geometry proofs with various visual effects. It could have many applications, e.g., in geometry education.

JGEX is still an ongoing developing system. The current version is *beta* 0.80 which is available in our website woody [2].

References

1. Cut-the-knot,
   http://www.cut-the-knot.org/pythagoras/index.shtml
2. Chou, S., Gao, X., Ye, Z.: Java geometry expert server (2009),
   http://woody.cs.wichita.edu
3. Chou, S., Gao, X., Zhang, J.: Automated generation of readable proofs with geometric invariants, II. Theorem proving with full-angles. Journal Automated Reasoning 17, 325–347 (1996)
4. Chou, S.C.: Mechanical geometry theorem proving. Springer, Heidelberg (1988)
5. Chou, S.C., Gao, X., Ye, Z.: Java Geometry Expert. In: Proceedings of the 10th Asian Technology Conference in Mathematics, pp. 78–84 (2005)
6. Chou, S.C., Gao, X.S., Zhang, J.Z.: A deductive database approach to automated geometry theorem proving and discovering. Journal of Automated Reasoning 25(3), 219–246 (2000)
7. Dolzmann, A., Sturm, T., Weispfenning, V.: A new approach for automatic theorem proving in real geometry. Journal of Automated Reasoning 21(3), 357–380 (1998)
8. Gao, X.S., Zhang, J.Z., Chou, S.C.: Geometry Expert. Nine Chapters Pub. (1998) (in Chinese)
9. Hongbo, L., Minde, C.: Proving theorems in elementary geometry with Glifford algebraic method. Chinese Math. Progress 26(4), 357–371 (1997)
10. Li, H.: Some applications of Clifford algebra to geometries. Automated Deduction in Geometry, 156–179 (1999)
11. Nelsen, R.: Proofs without words: Exercises in visual thinking. Mathematical Assn of Amer (1993)
12. Nelsen, R.: Proofs without words: More exercises in visual thinking. Mathematical Assn of Amer (2001)
13. Nelsen, R., Alsina, C.: Math Made Visual: Creating Images for Understanding Mathematics. Mathematical Assn of Amer (2006)
14. Wang, D.: Reasoning about geometric problems using an elimination method. Automatic Practical Reasoning, 147–185 (1989)

15. Wang, D.M., Gao, X.S.: Geometry theorems proved mechanically using Wu'method–part on Euclidean geometry. Mathematics-Mechanization Research Preprints 2 (1987)
16. Wen-Tsun, W.: On the decision problem and the mechanization of theorem proving in elementary geometry. Scientia Sinica 21(2), 159–172 (1978)
17. Ye, Z., Chou, S.C., Gao, X.S.: Visually Dynamic Presentation of Proofs in Plane Geometry Part 2. Automated Generation of Visually Dynamic Presentations with the Full-Angle Method and the Deductive Database Method. Journal of Automated Reasoning 45(3), 243–266 (2010)

On the Heilbronn Optimal Configuration of Seven Points in the Square[*]

Zhenbing Zeng and Liangyu Chen

Shanghai Key Laboratory of Trustworthy Computing
East China Normal University
200062 Shanghai, China
{zbzeng,lychen}@sei.ecnu.edu.cn

Abstract. In this paper, we prove that for any seven points in a unit square there exist three points whose area is not greater than a constant $h_7 = 0.083859...$ as conjectured by Francesc Comellas and J. Luis A. Yebra in 2002.

Keywords: Heilbronn Problem, Combinatorial Geometry, Automated Deduction.

1 Introduction

The Heilbronn problem for a given bounded closed convex set K in the plane and an integer $n \geq 3$ is to find the maximum of the following

$$h_{K,n}(p_1, p_2, \cdots, p_n) = \min\{\text{area}(p_i p_j p_k) | 1 \leq i < j < k \leq n\}$$

over all $p_1, p_2, \cdots, p_n \in K$. A configuration $\{p_1, p_2, \cdots, p_n\} \subset K$ is called a Heilbronn configuration of n points in K if it reaches the maximal value. Let $h(K, n)$ be the maximum corresponding to given K, n.

There has been a lot of work on the approximation of $h(K, n)$ for large n. The exact values of $h(K, n)$ are known only for $n \leq 6$ and K is square, triangle and disk. There are also some conjectures on the possible Heilbronn configurations obtained by numeric optimization methods (see [5,3,9]). The following is such a conjecture for 7 points in a square established by Francesc Comellas, J. Luis A. Yebra in [3].

Conjecture 1. *For any seven points $p_1, p_2, \cdots, p_7$ in the unit square S, the inequality*

$$\min\{\text{area}(p_i p_j p_k) | 1 \leq i < j < k \leq 7\} \leq \frac{2z^2 + 14z - 1}{38} \approx 0.083859 \cdots$$

---

[*] This work is supported by the National Natural Science Foundation of China (No. 10471044) and the Major Research Plan of the National Natural Science Foundation of China (No. 90718041).

T. Sturm and C. Zengler (Eds.): ADG 2008, LNAI 6301, pp. 196–224, 2011.
© Springer-Verlag Berlin Heidelberg 2011

holds, where z is the smallest positive real root of equation

$$z^3 + 5z^2 - 5z + 1 = 0, \quad z = 0.287258\cdots.$$

The equality holds for the configuration shown in Fig. 1.

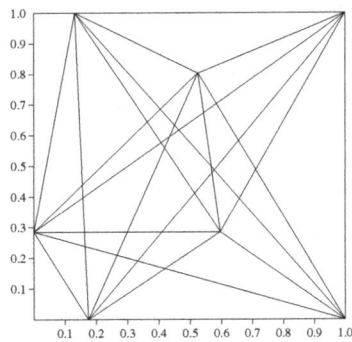

Fig. 1. The Heilbronn configuration of 7 points in the unit square

In this paper, we give a strict proof to this conjecture. The proof is composed of the following steps. The first step is to classify the possible optimal configurations according to the combinatorial type of the configuration and the allocation of the configuration in the unit square so that the original problem can be reduced to searching of finitely many types of local optimal configurations, and each of these local optimizations can be represented as a non-linear programming problem. In the second step we simplify each non-linear optimization problem through finding the loose constraints of the feasible set of the problem. In the third step we devote to get the upper bounds of the reduced non-linear programming problems and prove that among these problems there is only one case (in the sense of isometric transformation) can reach the global optimization and in all other local optimizations the smallest triangle is not greater than $1/12$ of the unit square.

2 Reducing the Optimal Configuration to 226 Local Optimizations

In this section we investigate the collocation of possible optimal configurations within the unit square and reduce the original problem to a finite many local optimization problems. This step is completed by the following 5 sub-steps.

Sub-Step 1. Prove that if a configuration of 7 points $p_1, p_2, \cdots, p_7 \in S$ is the optimal configuration of Heilbronn Problem for seven points, then the convex hull of the seven points can not be a triangle or a tetragon. This claim is relied on the following known property [4,7,9].

Lemma 1. *If points P_4, P_5 are contained in the triangle $P_1P_2P_3$, then the smallest triangles formed by the five points is less than $1/(4+2\sqrt{3}) < 1/6$ of the area of $P_1P_2P_3$.*

Sub-Step 2. Let Σ_k ($5 \leq k \leq 7$) be the set of all configurations of 7 points $p_1, p_2, \cdots, p_7 \in S$ such that its convex hull is a k-gon. Prove that if $\{p_1, p_2, \cdots, p_7\} \in \Sigma_k$ is a Heilbronn configuration, then there exists a parallelogram $ABCD$ of unit area that covers the convex hull (say, $P_1P_2\cdots P_k$) of the optimal configuration in one of the 13 forms shown in Fig.2, Fig.3 and Fig.4. This result is obtained by using of the following property proved in [8,9].

Lemma 2. *If $K = p_1p_2\cdots p_n$ is a convex n-gon and $ABCD$ is one of the parallelogram covering K with the smallest area, then each edge of $ABCD$ contains at least one of the vertices of K, and furthermore,*

1. *if K and $ABCD$ have no common vertex, then $\{p_1, p_2, \cdots, p_n\}$ has at least five points contained in the four edges of $ABCD$, and*
   - (a) *if $K \cap BC = \{p_k\}, K \cap DA = \{p_m\}$, then $p_kp_m//AB$,*
   - (b) *if $K \cap AB = p_ip_j(j = i+1), K \cap CD = \{p_l\}, p'_l = \ell(p_l, BC) \cap AB$, where $\ell(p_l, BC)$ the line that passes through p_l and parallels to BC, then p'_l is contained in the interior of the segment p_ip_j,*
   - (c) *if $K \cap AB = p_ip_j(j = i + 1), K \cap CD = p_lp_m$ and $p'_l = \ell(p_l, BC) \cap AB, p'_m = \ell(p_m, BC) \cap AB$, then $p'_lp'_m \cap p_ip_j \neq \emptyset$;*
2. *if K and $ABCD$ have one common vertex, say, $p_1 = A$, then in the interior of each edge of $ABCD$ there exists one or two vertices of K, say, p_2 contained in the interior of AB and p_n in the interior of DA, and*
   - (a) *if BC or CD contains only one vertex of K, say $K \cap BC = \{p_i\}$, then $p'_i \in p_np_1$, where $p'_i = \ell(p_i, AB) \cap DA$,*
   - (b) *if BC or CD contains two vertices of K, say $K \cap BC = p_ip_j, p_j \in p_iC$, then $p'_i \in p_np_1$ also holds, where $p'_i = \ell(p_i, AB) \cap DA$;*
3. *if K and $ABCD$ have two common vertices and these two points form a diagonal of $ABCD$, say, $p_1 = A, p_k = C$, then each edge of $ABCD$ also contains one vertex of K in its interior.*

Sub-Step 3. Then we prove that if the optimal configuration $\{p_1, p_2, \cdots, p_7\}$ is of category Σ_5, $P_1P_2\cdots P_5$ its convex hull, and $R_0, R_1, \cdots, R_9, R_X$ the 11 sub-regions of the convex hull divided the diagonals as shown in Fig.5, then $P_6, P_7 = \{p_1, p_2, \cdots, p_7\} \setminus \{P_1, P_2, \cdots, P_5\}$ satisfy one of the following properties

$$(P_6 \in R_9, P_7 \in R_2) \vee (P_6 \in R_9, P_7 \in R_6)$$
$$\vee (P_6 \in R_0, P_7 \in R_7) \vee (P_6 \in R_0, P_7 \in R_3)$$
$$\vee (P_6 \in R_8, P_7 \in R_1) \vee (P_6 \in R_8, P_7 \in R_5)$$

up to a permutation of P_6, P_7. This can be proved by using Lemma 1 and the following result (cf. [8]).

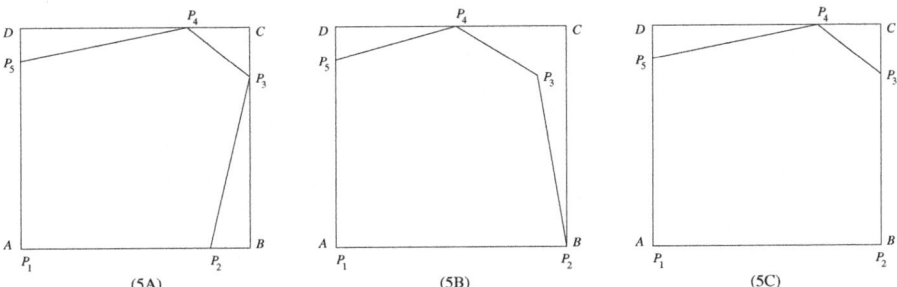

Fig. 2. The minimal parallelogram covering Σ_5 configurations. Note that $DP_4 \leq AP_2, BP_3 \leq AP_5$ in (5A) and (5C), and $DP_4 < AP_2, \text{area}(P_3AB) \leq \text{area}(P_5AB)$ in (5B).

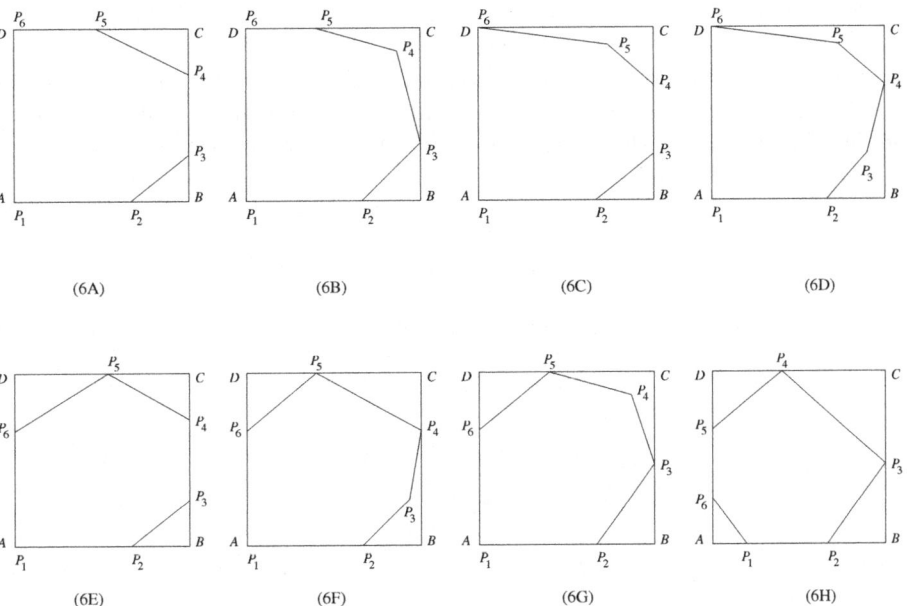

Fig. 3. The minimal parallelogram covering Σ_6 configurations. Note that $DP_5 \leq AP_2, BP_3 \leq AP_6$ in (6E), (6G), $DP_5 \leq AP_2, BP_4 \leq AP_6$ in (6F), and $AP_1 \leq DP_4 \leq AP_2, AP_6 \leq BP_3 \leq AP_5$ in (6H).

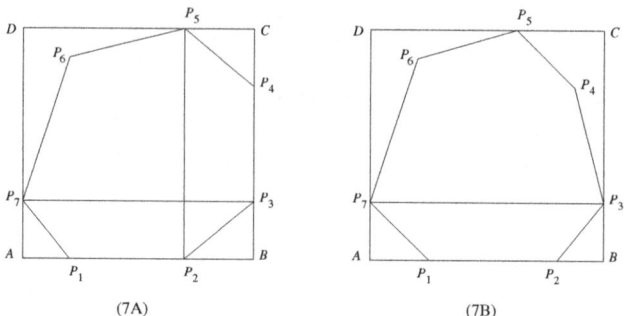

Fig. 4. The minimal parallelogram covering Σ_7 configurations. Note that $P_2P_5//BC$, $P_7P_3//AB$ in (7A), and $P_7P_3//AB$ in (7B).

Lemma 3. *In the configuration (5A) shown in Fig.2, $P_1P_2 > DP4, P_1P_5 > BP_3$, then*

$$\text{area}(P_1P_4P_5) + \text{area}(P_2P_3P_4) < \frac{1}{2}, \quad \text{area}(P_1P_2P_3) + \text{area}(P_3P_4P_5) < \frac{1}{2}.$$

Using this property we deduce the computation of the possible Σ_5-category optimal configurations into $3 \times 6 = 18$ non-linear optimization problems with 8 (for (5C) in Fig.2) to 9 (for (5A) and (5B) in Fig.2) unknowns.

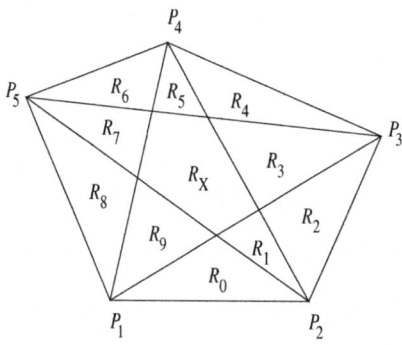

Fig. 5. The sub-regions divided by the diagonals of a pentagon

Sub-Step 4. For the possible Σ_6 optimal configuration, each type in Fig.3 can be decomposed into 26 sub-types (explained later) according to the position the point (say, p_7) that is not a vertex of the convex hull (see Fig.6). This leads the computation of the Σ_6 optimization configurations to $8 \times 26 = 208$ (the number can be reduced if consider the symmetry in (6A) and (6H) configurations in

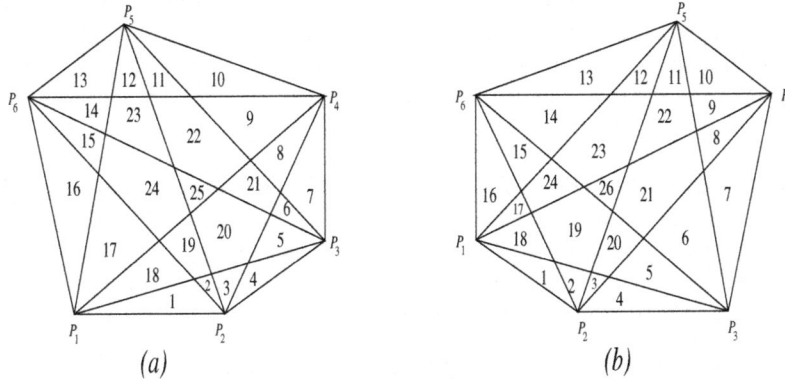

Fig. 6. The oriented sub-regions of a convex hexagon divided by its diagonals

Fig.3) non-linear optimization problems with 7 (for configuration (6A)) to 9 (for (6D), (6F), (6G) and (6H) in Fig.3) unknowns.

To explain the 26 sub-types of Σ_6, we need to use the concept of orientation of triangles. Let ABC be any triangle with $A = (x_1, x_2)$, $B = (x_3, x_4)$, $C = (x_5, x_6)$. Then the oriented area of ABC is

$$\text{area}(ABC) = \frac{1}{2} \cdot \begin{vmatrix} x_1 & x_2 & 1 \\ x_3 & x_4 & 1 \\ x_5 & x_6 & 1 \end{vmatrix}.$$

We will call that the orientation of ABC is positive (or orient$(ABC) = 1$) if $S(ABC) > 0$, and the orientation is negative (or orient$(ABC) = -1$) if $S(ABC) < 0$, as shown in Fig. 7. We may call a triangle ABC is degenerate if area$(ABC) = 0$.

Note that the diagonals of a convex hexagon divide the hexagon into 24 or 25 small regions. Each region is a triangle, or a quadrilateral, or a pentagon. If the three major diagonals meet at the same point, then the number of the small regions is 24. Otherwise it is 25. Let Δ be the triangle region formed by the three major diagonals. Then according to the orientation of Δ the shapes

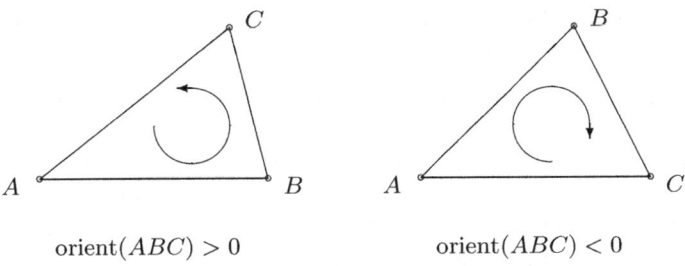

Fig. 7. Two triangles with different orientations

of 24 small regions can be classified into two cases, as shown in Fig.6. In Case (a), regions with number 19, 21, 23 are quadrilateral, regions 20, 22, 24 are pentagon. In Case (b), regions 19, 21, 23 are pentagon and regions 20, 22, 24 are quadrilateral. In both cases, regions with number $1, 2, \cdots, 18$ are all triangles. The following observations are obviously true.

Lemma 4. *Let $P_1 P_2 \cdots P_6$ and $P'_1 P'_2 \cdots P'_6$ be the two hexagons corresponding to Case (a) and Case (b) in Fig. 6. Then for any natural number k with $1 \leq k \leq 24$, any point Q that is contained in the region with number k in Case (a), and any point Q' that is contained in the region with number k in Case (b), the orientation of triangle $QP_i P_j$ is same as the that of $Q' P'_i P'_j$ for all i, j with $1 \leq i < j \leq 6$.*

Lemma 5. *Let $P_1 P_2 \cdots P_6$ and $P'_1 P'_2 \cdots P'_6$ be the two hexagons corresponding to Case (a) and Case (b) in Fig. 6. Then for any point Q that is contained in the triangle formed by the three major diagonals (region with number 25) in Case (a), and any point Q' that is contained in the triangle formed by the three major diagonals (region with number 26) in Case (b), the orientation of triangle $QP_i P_j$ and the orientations of $Q' P'_i P'_j$ differ only for the following cases.*

$$\text{orient}(QP_1 P_4) = -\text{orient}(Q' P'_1 P'_4) = 1;$$
$$\text{orient}(QP_2 P_5) = -\text{orient}(Q' P'_2 P'_5) = -1;$$
$$\text{orient}(QP_3 P_6) = -\text{orient}(Q' P'_3 P'_6) = 1.$$

Sub-Step 5. For the two categories defined in Fig.4, that is, the categories corresponding to that optimal configuration $\{p_1, p_2, \cdots, p_7\}$ is a convex heptagon, it is easy to prove that the (local) optimal configuration satisfies the following property (cf. [9,6])

$$\text{area}(p_1 p_2 p_3) = \text{area}(p_2 p_3 p_4) = \cdots = \text{area}(p_7 p_1 p_2),$$

and therefore, $p_1 p_2 \cdots p_7$ is the maximal regular heptagon inscribed in the unit square,

$$\min\{\text{area}(p_i p_j p_k) : 1 \leq i < j < k \leq 7\} = \frac{\sin^2(2\pi/7) \tan(\pi/7)}{2(1 + \cos(\pi/7)) \sin(3\pi/7)} < \frac{1}{12},$$

which shows that the local optimal configurations in Σ_7 can not be the global optimization.

To summarize this step we known that $p_1 p_2 \cdots p_7$ is the Heilbronn optimal configuration of seven points in the unit square, then either the convex hull of $p_1 p_2 \cdots p_7$ is a convex pentagon, say, $P_1 P_2 \cdots P_5$, satisfying that the configuration $\{P_1, P_2, \cdots, P_5, A, B, C, D\}$ is belong to one of the three categories $(5A), (5B), (5C)$ and $P_6, P_7 = \{p_1, p_2, \cdots, p_7\} \setminus \{P_1, P_2, \cdots, P_5\}$ are contained in one of the following six 2-tuples of regions:

$$(R_9, R_2), \ (R_9, R_6), \ (R_0, R_7), \ (R_0, R_3), \ (R_8, R_1), \ (R_8, R_5);$$

or the convex hull of $p_1p_2 \cdots p_7$ is a convex hexagon, say, $P_1 P_2 \cdots P_6$, satisfying that the configuration of $\{P_1, P_2, \cdots, P_6, A, B, C, D\}$ is belong to one of the eight categories $(6A), (6B), \cdots, (6H)$ and the point $P_7 = \{p_1, p_2, \cdots, p_7\} \setminus \{P_1, P_2, \cdots, P_6\}$ is contained in one of the 26 oriented sub-regions divided by the diagonals of the hexagon. This reduced the original problem to $6 \times 3 + 26 \times 8 = 226$ sub-problems with less freedoms which can also be regarded as local optimization problems.

3   Checking the Loose Constraints of Non-linear Programming Problems

The second part of the proof is to simplify the constraints of the $18 + 208 = 226$ non-linear optimization problems. Each problem is in the following form

$$\max x_0,$$
$$\text{subject to}$$
$$S_{i,j,k}(x_1, \cdots, x_q) \geq x_0, (1 \leq i < j < k \leq 7)$$
$$0 \leq x_1, \cdots, x_q \leq 1.$$

where $S_{i,j,k}(x_1, x_2, \cdots, x_q)$ are quadratic polynomials determined by the area formula

$$\text{area}(p_i p_j p_k) = \frac{1}{2} \cdot \begin{vmatrix} x_1 & x_2 & 1 \\ x_3 & x_4 & 1 \\ x_5 & x_6 & 1 \end{vmatrix}, \quad S_{i,j,k} = \text{orient}(p_i p_j p_k) \cdot \text{area}(p_i p_j p_k)$$

for oriented triangle formed by $p_i = (x_1, x_2), p_j = (x_3, x_4), p_k = (x_5, x_6)$. For simplicity we may assume that the coordinates of the seven points are

$$P_1 = (x_1, x_2), P_2 = (x_3, x_4), \cdots, \cdots, P_6 = (x_{11}, x_{12}), P_7 = (x_{13}, x_{14})$$

and we will use $f_1 \geq 0, f_2 \geq 0, \cdots, f_{35} \geq 0$ to represent the 35 constraints related to the area of triangles in the following way:

$$f_1 = f_1(x_0, x_1, \cdots, x_{14}) = S_{1,2,3} - x_0,$$
$$f_2 = f_2(x_0, x_1, \cdots, x_{14}) = S_{1,2,4} - x_0,$$
$$f_3 = f_3(x_0, x_1, \cdots, x_{14}) = S_{1,2,5} - x_0,$$
$$\cdots \cdots,$$
$$f_{34} = f_{34}(x_0, x_1, \cdots, x_{14}) = S_{4,6,7} - x_0,$$
$$f_{35} = f_{35}(x_0, x_1, \cdots, x_{14}) = S_{5,6,7} - x_0.$$

We shall use $(5Xij), (5Xij), (5Xij)$ where $X \in \{A, B, C\}$ and

$$(i, j) \in \{(9, 2), (9, 6), (0, 7), (0, 3), (8, 1), (8, 5)\}$$

to denote the 18 non-linear programming problems corresponding to the possible Σ_5 optimal configurations and use $(6Xk)$ where $X \in \{A, B, \cdots, H\}$ and $1 \le k \le 26$ to denote the 208 non-linear programming problems corresponding to the possible Σ_6 optimal configurations.

For the linear constraint part $0 \le x_i \le 1(i = 1, 2, \cdots, 14)$, we have the following known coordinates for $(5Xij)$ problems:

$$(5Aij) : x_1 = 0, \quad x_2 = 0, \quad x_4 = 0, \quad x_5 = 1, \quad x_8 = 1, \quad x_9 = 0;$$
$$(5Bij) : x_1 = 0, \quad x_2 = 0, \quad x_3 = 1, \quad x_4 = 0, \quad x_8 = 1, \quad x_9 = 0;$$
$$(5Cij) : x_1 = 0, \quad x_2 = 0, \quad x_3 = 1, \quad x_4 = 0, \quad x_5 = 1, \quad x_8 = 1, \quad x_9 = 0;$$

and the following known for $(6Xk)$ problems:

$$(6Ak) : x_1 = 0, \quad x_2 = 0, \quad x_4 = 0, \quad x_5 = 1, \quad x_7 = 1,$$
$$x_{10} = 1, \quad x_{11} = 0, \quad x_{12} = 1;$$
$$(6Bk) : x_1 = 0, \quad x_2 = 0, \quad x_4 = 0, \quad x_5 = 1,$$
$$x_{10} = 1, \quad x_{11} = 0, \quad x_{12} = 1;$$
$$(6Ck) : x_1 = 0, \quad x_2 = 0, \quad x_4 = 0, \quad x_5 = 1, \quad x_7 = 1,$$
$$x_{11} = 0, \quad x_{12} = 1;$$
$$(6Dk) : x_1 = 0, \quad x_2 = 0, \quad x_4 = 0, \quad x_7 = 1,$$
$$x_{11} = 0, \quad x_{12} = 1;$$
$$(6Ek) : x_1 = 0, \quad x_2 = 0, \quad x_4 = 0, \quad x_5 = 1, \quad x_7 = 1,$$
$$x_{10} = 1, \quad x_{11} = 0;$$
$$(6Fk) : x_1 = 0, \quad x_2 = 0, \quad x_4 = 0, \quad x_7 = 1,$$
$$x_{10} = 1, \quad x_{11} = 0;$$
$$(6Gk) : x_1 = 0, \quad x_2 = 0, \quad x_4 = 0, \quad x_5 = 1,$$
$$x_{10} = 1, \quad x_{11} = 0;$$
$$(6Hk) : x_2 = 0, \quad x_4 = 0, \quad x_5 = 1, \quad x_8 = 1,$$
$$x_9 = 0, \quad x_{11} = 0.$$

And all other coordinates x_n in non-linear programming problems $(5Xij), (6Xk)$ satisfy the strict inequality $0 < x_n < 1$.

Note that if we take $x_3 = 1$ in $(5Aij)$ or take $x_5 = 1$ in $(5Bij)$ then the problems changed to $(5Cij)$. That is, $(5Cij)$ is a degenerate case of $(5Aij)$ or $(5Bij)$. This can be shown by the following diagram.

$$(5Aij) \longrightarrow (5Cij) \longleftarrow (5Bij)$$

The following diagram shows the degenerate relations among non-linear programming problems $(6Ak), (6Bk), \cdots, (6Gk)$.

$$(6Ak)$$

$$(6Dk) \longrightarrow (6Ck) \longrightarrow (6Bk) \qquad\qquad (6Ek) \longleftarrow (6Fk)$$

$$(6Gk)$$

The following simple observation can be used to simplify the constraints of the non-linear optimizations.

Lemma 6. *Let* Area$(\cdot)$ *be the area of a polygon.*
(i) If p_4 is contained in the interior of a triangle $p_1p_2p_3$, then

$$\text{Area}(p_1p_2p_3) > \min\{\text{Area}(p_ip_jp_k)|1 \le i \le 4\}.$$

(ii) If $p_1p_2 \cdots p_5$ is a convex pentagon, then

$$\text{Area}(p_1p_2p_4) > \min\{\text{Area}(p_ip_jp_k)|1 \le i \le 5\}.$$

With this property we can find a subset S of $\{f_1, f_2, \cdots, f_{35}\}$ such that if $(x_0, x_1, \cdots, x_{14})$ is any feasible solution of $(5Xij)$ or $(6Xk)$, then all member $f \in S$ satisfies the strict inequality $f(x_0, x_1, \cdots, x_{14}) > 0$. We show how to do this for $(5X92)$ and $(6X16)$ as examples.

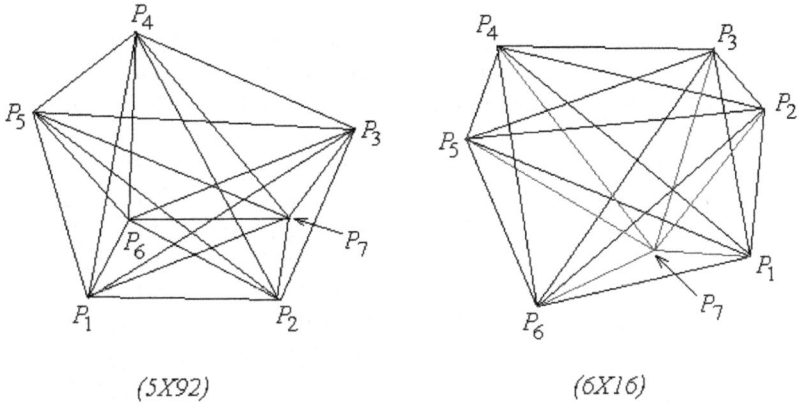

(5X92)   (6X16)

Fig. 8. Configurations corresponding to $(5X92)$ and $(6X16)$

In $(5X92)$ configurations, we have

$$(1) \qquad \begin{array}{l} P_6 \in P_1P_2P_4, P_1P_2P_5, P_1P_3P_4, P_1P_3P_5, P_1P_4P_7, P_1P_5P_7, \\ P_7 \in P_1P_2P_3, P_2P_3P_4, P_2P_3P_5, P_2P_3P_6, \end{array}$$

and the following 6 convex pentagons formed by points $P_1, P_2, \cdots, P_7$

$$P_1P_2P_3P_4P_5, P_1P_2P_7P_4P_5, P_1P_7P_3P_4P_5,$$
$$P_2P_3P_4P_5P_6, P_2P_7P_4P_5P_6, P_3P_4P_5P_6P_7,$$

from which we know that the following triangles are not the smallest ones:

(2)
$$\text{in } P_1P_2P_3P_4P_5 : P_1P_2P_4, P_2P_3P_5, P_1P_3P_4, P_2P_4P_5(*), P_1P_3P_5;$$
$$\text{in } P_1P_2P_7P_4P_5 : P_1P_2P_4, P_2P_5P_7(*), P_1P_4P_7, P_2P_4P_5, P_1P_5P_7;$$
$$\text{in } P_1P_7P_3P_4P_5 : P_1P_4P_7, P_3P_5P_7(*), P_1P_3P_4, P_4P_5P_7(*), P_1P_5P_7;$$
$$\text{in } P_2P_3P_4P_5P_6 : P_2P_3P_5, P_3P_4P_6(*), P_2P_4P_5, P_3P_5P_6(*), P_2P_4P_6(*);$$
$$\text{in } P_2P_7P_4P_5P_6 : P_2P_5P_7, P_4P_6P_7(*), P_2P_4P_5, P_5P_6P_7(*), P_2P_4P_6;$$
$$\text{in } P_3P_4P_5P_6P_7 : P_3P_4P_6, P_4P_5P_7, P_3P_5P_6, P_4P_6P_7, P_3P_5P_7;$$

where the $(*)$ after a triangle is to show that the triangle is neither contained in (1) nor previous lines of (2). So we have

$$S(5X92) = \{f_1, f_2, f_3, f_6, f_7, f_{12}, f_{14}, f_{16}, f_{17}, f_{18},$$
$$f_{20}, f_{21}, f_{24}, f_{26}, f_{28}, f_{29}, f_{31}, f_{33}, f_{35}\}$$

for $X = A, B, C$. In $(6X16)$ configurations we have

$$(3) \qquad P_7 \in P_1P_2P_6, P_1P_3P_6, P_1P_4P_6, P_1P_5P_6,$$

and only the following 5 pentagons

$$P_1P_2P_4P_6P_7, P_1P_2P_5P_6P_7, P_1P_3P_4P_6P_7, P_1P_3P_5P_6P_7, P_1P_4P_5P_6P_7$$

formed by $P_1, P_2, \cdots, P_7$ are not convex. In the following list we use $P_iP_jP_k(mn)$ to denote that $\{P_i, P_j, P_k, P_m, P_n\}$ formed a convex pentagon(and therefore, $P_iP_jP_k$ is not the smallest one among triangles formed by these five points).

(4)
$$P_1P_2P_3, P_1P_2P_4(36), P_1P_2P_5(36), P_1P_2P_6, \quad P_1P_2P_7,$$
$$P_1P_3P_4(26), P_1P_3P_5(46), P_1P_3P_6, \quad P_1P_3P_7(25),$$
$$P_1P_4P_5(36), P_1P_4P_6(25), P_1P_4P_7(25),$$
$$P_1P_5P_6, \quad P_1P_5P_7,$$
$$P_1P_6P_7,$$
$$P_2P_3P_4, \quad P_2P_3P_5(46), P_2P_3P_6(41), P_2P_3P_7(41),$$
$$P_2P_4P_5(36), P_2P_4P_6(51), P_2P_4P_7(51),$$
$$P_2P_5P_6(41), P_2P_5P_7(36),$$
$$P_2P_6P_7,$$
$$P_3P_4P_5, \quad P_3P_4P_6(25), P_3P_4P_7(25),$$
$$P_3P_5P_6(41), P_3P_5P_7(41),$$
$$P_3P_6P_7(52),$$
$$P_4P_5P_6, \quad P_4P_5P_7(36),$$
$$P_4P_6P_7(52),$$
$$P_5P_6P_7.$$

Thus

$$S(6X16) = \{f_2, f_3, f_4, f_6, f_7, f_8, f_9, f_{10}, f_{11}, f_{12}, f_{13}, f_{17}, f_{18}, f_{19},$$
$$f_{20}, f_{21}, f_{22}, f_{23}, f_{24}, f_{27}, f_{28}, f_{29}, f_{30}, f_{31}, f_{33}, f_{34}\}$$

has 26 members for $X \in \{A, B, \cdots, H\}$. This means, the non-linear optimization problems (6A16) can be simplified to the following one:

(6A16)
$$\begin{aligned}
\max\ &x_0, \\
\text{s.t.}\ &f_1 \geq 0, f_5 \geq 0, f_{14} \geq 0, f_{15} \geq 0, f_{16} \geq 0, \\
&f_{25} \geq 0, f_{26} \geq 0, f_{32} \geq 0, f_{35} \geq 0; \\
&x_1 = 0, x_2 = 0, x_4 = 0, x_5 = 1, x_7 = 1, \\
&x_{10} = 1, x_{11} = 0, x_{12} = 1; \\
&0 < x_i < 1 (i = 3, 6, 8, 9, 13, 14).
\end{aligned}$$

The above procedure for simplifying the constraints can be realized automatically by a computer program. The following is the results for all non-linear programming problems (5Xij) and (6Xk).

Theorem 1. Let $\#(\cdot)$ be the cardinal of a set. Then

$$\#S(5Xij) = 19$$

for all $X = A, B, C$ and $(i, j) \in \{(9, 2), (9, 6), (0, 7), (0, 3), (8, 1), (8, 5)\}$, and

$$\begin{aligned}
\#(6Xk) &= 26(k = 1, 4, 7, 10, 13, 16), \\
\#(6Xk) &= 24(k = 2, 3, 5, 6, 8, 9, 11, 12, 14, 15, 17, 18), \\
\#(6Xk) &= 21(k = 19, 20, 21, 22, 23, 24), \\
\#(6X25) &= \#(6X26) = 23,
\end{aligned}$$

for all $X = A, B, \cdots, H$.

Proof. The proof is obtained by applying Lemma 6 to the corresponding 226 configurations.   Q.E.D.

So we have reduced each of the 226 non-linear optimization problems to a problem with less unknowns and less constraints, namely, to maximize the linear function $F(y_0, y_1, \cdots, y_q) = y_0$ ($q \leq 8$), where $y_1, y_2, \cdots, y_q \in \{x_1, x_2, \cdots, x_{14}\}$, over a bounded subset $Q \subset R^{q+1}$ formed by $p = 16$ or less quadratic (or linear) polynomials of $y_0, y_1, y_2, \cdots, y_q$ and the following linear inequalities

$$0 < y_0 < \frac{1}{9}, \quad 0 < y_i < 1 (i = 1, 2, \cdots, q).$$

Here the upper bound $\frac{1}{9}$ of y_0 is based on the fact that if a pentagon or hexagon K is contained in the unit square then the area of K is less than 1, and the following known result (cf. [6]).

Lemma 7. For any seven points $P_1, P_2, \cdots, P_7$ in the plane, there exists a triangle formed by the points whose area is not more than 1/9 of the area of the convex hull of the seven points.

For computing the global maximum of the 226 problems, we may restrict each problem to $x_0 > 1/12$. Since the configuration found by Francesc Comellas, J. Luis A. Yebra shows that the optimal for Σ_5 satisfies

$$h_{\Sigma_5} \geq 0.083859 \cdots = \frac{1}{11.92477 \cdots} > \frac{1}{12},$$

and for Σ_6 configurations a known result (see [9]) is

$$h|_{\Sigma_6} \geq \frac{1}{12}$$

as shown in the following figure.

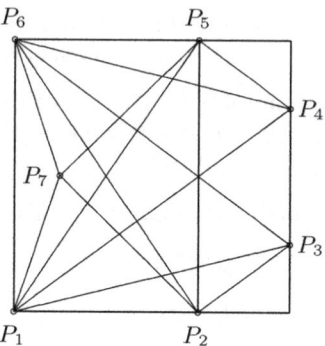

Fig. 9. A possible optimal configuration in Σ_6

4 Solving Non-linear Programming With Symbolic Computation

The third part of the proof is to solve the 226 non-linear programming problems. We may define an order on these problems, say,

$$(5C92) \prec (5C96) \prec (5C07) \prec (5C03) \prec (5C81) \prec (5C85)$$
$$\prec (5B92) \prec \cdots \prec (5B85) \prec (5A92) \prec \cdots \prec (5A85)$$
$$\prec (6A1) \prec (6A2) \prec \cdots \prec (6A26) \prec \cdots \prec (6H26),$$

and use the notation $SP(n)$ $(1 \leq n \leq 226)$ to represent the n-th sub-problem in this sequence. It is clear now that $SP(n)$ is in the following general form

$$\text{Sub-Problem } SP(n):$$
$$\max y_0;$$
$$\text{s.t. } g_1(y_1, \cdots, y_q) - y_0 \geq 0,$$
$$\cdots,$$
$$g_p(y_1, \cdots, y_q) - y_0 \geq 0,$$
$$0 < y_0 < \frac{1}{9},\ 0 < y_1 < 1, \cdots, 0 < y_q < 1,$$

Note that for each problem $SP(n)$, the point set $Q(n) \subset R^{q+1}$ defined by the constraints,

$$Q(n) = \{(y_0, y_1, \cdots, y_q)|0 < y_0 < 1/9, 0 < y_i < 1 \text{ for all } i = 1, \cdots, q,$$
$$g_j(y_1, \cdots, y_q) - y_0 \geq 0 \text{ for all } j = 1, \cdots, p\},$$

is non-empty and bounded. Let

$$Q(n, a) = \{(y_1, \cdots, y_q) | 0 < y_i < 1 \text{ for all } i = 1, \cdots, q,$$
$$g_j(y_1, \cdots, y_q) - a \geq 0 \text{ for all } j = 1, \cdots, p\} \subset R^q$$

be defined for all real number a. Then $Q(n, a)$ can be considered as (a projection of) the intersection of $Q(n)$ and the hyperplane $x_0 = a$. It is clear that

$$Q(n, a_2) \subset Q(n, a_1) \text{ if } a_1 < a_2,$$

and the optimal solution of $SP(n)$ is the maximal value a such that $Q(n, a)$ is non-empty, that is, if a is the optimal solution of $SP(n)$ then $Q(n, a) \neq \emptyset$ and $Q(n, a') = \emptyset$ for all $a' > a$. It is possible that a sub-problem $SP(n)$ has no solution since the constraint $Q(n)$ may not be a closed subset, that is, there exists a sequence $a_1, a_2, \cdots, a_k, \cdots$ of real numbers such that

$$0 < a_1 < a_2 < \cdots < a_k < \cdots < 1/9, \quad \lim_{k \to \infty} a_k = a_0,$$

and all sets $Q(n, a_k)$ are non-empty, but $Q(n, a_0)$ is empty.

The following property shows that for computing the global maximum of the 226 non-linear programming problems, we only need to consider that problems $SP(n)$ for which $SP(n)$ has solutions.

Lemma 8. *If $SP(n)$ has no solution, then there exists n' with $1 \leq n' \leq n$ such that*

$$Q(n, a) \neq \emptyset \Longrightarrow Q(n', a) \neq \emptyset$$

for all real number $a > 0$ and $SP(n')$ has a solution.

Proof. It is clear that for each sub-problem $SP(n)$, there exists at least a configuration, namely, a specified set of $x_1, x_2, \cdots, x_{14}$ such that the $p_1 = (x_1, x_2), \cdots, p_7 = (x_{13}, x_{14})$ satisfy $a = \min S_{i,j,k}(p_i p_j p_k) > 0$. For this specified a we have $Q(n, a) \neq \emptyset$. Let

$$a_0 = \sup\{a > 0 | Q(n, a) \neq \emptyset\},$$

and $a_1, a_2, \cdots, a_k, \cdots$ a sequence of real numbers such that

$$Q(n, a_k) \neq \emptyset \text{ for } k = 1, 2, \cdots, \quad \lim_{k \to \infty} a_k = a_0.$$

Then for any sequence $Y_k = (y_1^{(1)}, \cdots, y_q^{(1)}) \in Q(n, a_k), k = 1, 2, \cdots$, there is a subsequence Y_{k_l} such that

$$\lim_{l \to \infty} Y_{k_l} = (y_1^{(0)}, \cdots, y_q^{(0)})$$

exists. It is clear that

$$Y_0 := (y_1^{(0)}, \cdots, y_q^{(0)}) \in \overline{Q(n, a_0)} \setminus Q(n, a_0),$$

which implies that there exists at least one i with $1 \leq i \leq q$ such that $y_i^{(0)} = 0$. Meanwhile, Y_0 satisfies

$$g_j(y_1^{(0)}, \cdots, y_q^{(0)}) \geq a_0 > 0 \text{ for all } j = 1, 2, \cdots, p,$$

that means the configuration defined by Y_0 is in generic position, that is, no three points are collinear, and therefore, it is a degenerate configuration of $SP(n)$. According to the definition of the order on the 226 problems and the degenerate relations diagrams, this degenerate problem $SP(n')$ satisfies $n' < n$. This proves Lemma 8.   Q.E.D.

Given any n with $1 \leq n \leq 226$, if the sub-problem $SP(n)$ has a solution $y_0 = a$ and $(y_1, \cdots, y_q) \in S(n, a)$, then it is obvious that the set of indices of the tight constraints of $SP(n)$ defined by

$$T(n) := \{j | 1 \leq j \leq p, \ g_j(y_1, \cdots, y_q) - a = 0\}$$

is not empty and a is the maximal real number such that the following semi-algebraic system

$$(n, T(n), a) : \quad \begin{cases} g_i(y_1, \cdots, y_q) = a \text{ for } i \in T(n), \\ g_j(y_1, \cdots, y_q) > a \text{ for } j \notin T(n), \\ 0 < y_1 < 1, \cdots, 0 < y_q < 1, \end{cases}$$

has real solutions. For short we call this $T(n)$ the optimal tight constraints of $SP(n)$. In general, for each subset $T \subset \{1, 2, \cdots, p\}$ we can construct the following non-linear programming problem induced by the sub-problem $SP(n)$ and T,

$$NLP(n, T) :$$
$$\max y_0,$$
$$\text{s.t. } g_i(y1, \cdots, y_q) - y_0 = 0, \text{ for all } i \in T,$$
$$g_j(y_1, \cdots, y_q) - y_0 > 0, \text{ for all } j \notin T,$$
$$0 < y_0 < 1/9, 0 < y_1 < 1, \cdots, 0 < y_q < 1.$$

Then the sub-problem $SP(n)$ can be decomposed into a family of non-linear programming problems $NLP(n, T), T \in \{1, 2, \cdots, p\}$ in this form, and the optimal solution of $SP(n)$ is the maximal of solutions of $NLP(n, T)$ over all T.

It is also clear that for any sub-problem $SP(n)$ ($1 \leq n \leq 226$) and any $T_1, T_2 \subset \{1, 2, \cdots, p\}$ with $T_1 \subset T_2$, if a_1 and a_2 are solutions of $NLP(n, T_1)$ and $NLP(n, T_2)$, with respectively, then $a_1 \geq a_2$. In particularly, if the sub-problem $SP(n)$ has a solution $y_0 = a$ and $T(n)$ is the optimal tight constraints of $SP(n)$, then for subset $T \subset \{1, 2, \cdots, p\}$ with $T(n) \subset T$ and $T(n) \neq T$, either the problem $NLP(n, T)$ has no solution, or any real point $(a', y_1', \cdots, y_q')$ in the following semi-algebraic system

$$(n, T, h) : \quad \begin{cases} g_i(y_1, \cdots, y_q) = h \text{ for } i \in T, \\ g_j(y_1, \cdots, y_q) > h \text{ for } j \notin T, \\ 0 < y_1 < 1, \cdots, 0 < y_q < 1. \end{cases}$$

satisfies that $a' < a$.

The general idea for solving each non-linear programming problem $SP(n)$

$$\max y_0;$$
$$\text{s.t. } g_1(y_1, \cdots, y_q) - y_0 \geq 0, \cdots, g_p(y_1, \cdots, y_q) - y_0 \geq 0,$$
$$y_0 < \frac{1}{9}, \ 0 < y_1 < 1, \cdots, 0 < y_q < 1,$$

is to investigate each non-linear programming problem $NLP(n,T)$ to see if there exists a subset T of $\{1, 2, \cdots, p\}$ such that the corresponding problem $NLP(n,T)$ has a solution $y_0(n,T)$ which is better than the known result. Let

$$S = S(n) = \{T \mid T \subset \{1, 2, \cdots, p\}\}.$$

Since we are looking for the global maximum of $y_0(n,T)$ for $n = 1, 2, \cdots, 226$ and T over corresponding $S(n)$, and we know already that this maximum is not less than $1/12$ from the configuration in the Fig. 9, thus the sub-problem $SP(n)$ can be reduced to solving the family of the following modified problems,

$$NLP(n,T):$$
$$\max y_0,$$
$$\text{s.t. } g_i(y1, \cdots, y_q) - y_0 = 0, \text{ for all } i \in T,$$
$$g_j(y_1, \cdots, y_q) - y_0 > 0, \text{ for all } j \notin T,$$
$$1/12 < y_0 < 1/9, \ 0 < y_1 < 1, \cdots, 0 < y_q < 1,$$
$$(T \in S(n))$$

By $NLP(n,T) \neq \emptyset$ we denote that the modified problem $NLP(n,T)$ has a solution $y_0(n,T)$ such that $y_0(n,T) > 1/12$. Let

$$\mathcal{D}(n) = \{T \subset S \mid NLP(n,T) \neq \emptyset\}.$$

and

$$\mathcal{A}(n) = \{T \subset S \mid NLP(n,T) = \emptyset\},$$

Then our goal is to find for what n the set $\mathcal{D}(n)$ is not empty. For any $T \in S$ and $\mathcal{U} \subset S$ let $\text{mov}(T, \mathcal{U})$ be the following operation

$$\mathcal{U} := \mathcal{U} \cup \{T\}, \quad S := S \setminus \{T\},$$

and $\text{mov}^*(T, \mathcal{U})$ and $\text{mov}_*(T, \mathcal{U})$ be the operations defined by

$$\mathcal{V} := \{T_1 \mid T_1 \in S, T \subset T_1\};$$
$$\mathcal{U} := \mathcal{U} \cup \mathcal{V};$$
$$S := S \setminus \mathcal{V};$$

and

$$\mathcal{W} := \{T_1 \mid T_1 \in S, T_1 \subset T\};$$
$$\mathcal{U} := \mathcal{U} \cup \mathcal{W};$$
$$S := S \setminus \mathcal{W};$$

with respectively. Then the algorithm for checking if $\mathcal{D}(n)$ is not empty can be described in the following form.

Main Algorithm
```
 1. Initializing: Construct S. Let A = ∅, B = ∅, C = ∅, D = ∅,
 2. For all T ∈ S, do procedure (I) in the Sketch Diagram,
 3. For all T ∈ B, do procedure (II) in the Sketch Diagram,
 4. For all T ∈ C, do procedure (III) in the Sketch Diagram,
 5. Output D(n).
```

The following `Sketch Diagram` shows the three procedures (I), (II), (III) for doing operations $\mathrm{mov}(T, \mathcal{D})$, $\mathrm{mov}^*(T, \mathcal{A})$, $\mathrm{mov}_*(T, \mathcal{B})$, and $\mathrm{mov}_*(T, \mathcal{C})$ by using symbolic computation $S1, S2, \cdots, S7$.

$$(\mathrm{I}): \quad (n, T) \xrightarrow{S1} \begin{cases} (1): \mathrm{mov}^*(T, \mathcal{A}) & ① \\ (2): \xrightarrow{S2} \begin{cases} (21): \xrightarrow{S3} \begin{cases} (211): \mathrm{mov}^*(T, \mathcal{A}) & ② \\ (212): \mathrm{mov}(T, \mathcal{D}) & ③ \end{cases} \\ (22): \mathrm{mov}_*(T, \mathcal{C}) \dashrightarrow (\mathrm{III}) & ④ \end{cases} \\ (3): \mathrm{mov}_*(T, \mathcal{B}) \dashrightarrow (\mathrm{II}) & ⑤ \end{cases}$$

$$(\mathrm{II}): \quad (n, T)|_{T \in \mathcal{B}(n)} \xrightarrow{S4} \begin{cases} (31): \mathrm{mov}^*(T, \mathcal{A}) & ⑥ \\ (32): \xrightarrow{S5} \begin{cases} (321): \mathrm{mov}^*(T, \mathcal{A}) & ⑦ \\ (322): \mathrm{mov}(T, \mathcal{D}) & ⑧ \end{cases} \\ (33): \mathrm{mov}_*(T, \mathcal{C}) \dashrightarrow (\mathrm{III}) & ⑨ \end{cases}$$

$$(\mathrm{III}): (n, T)|_{T \in \mathcal{C}(n)} \xrightarrow{S6} \begin{cases} (41): \mathrm{mov}^*(T, \mathcal{A}) & ⑩ \\ (42): \xrightarrow{S7} \begin{cases} (421): \mathrm{mov}^*(T, \mathcal{A}) & ⑪ \\ (422): \mathrm{mov}(T, \mathcal{D}) & ⑫ \end{cases} \end{cases}$$

Here $(1), (2), \cdots, (422)$ represent the cases of the results obtained by the symbolic computation $S1, S2, \cdots, S7$. Thus, the procedure (I) can be transformed to the following form.

Procedure (I)
```
For T ∈ S(n) do symbolic computation S1.
   if the result belong to Case (1), then mov*(T, A),
   else if the result belong to Case (2), then S2,
     if this result belong to Case (21), then do S3,
       if this new result belong to Case (212), then mov*(T, A),
```

```
    else mov(T,D),
   else mov*(T,C) and go to Procedure (III),
 else mov*(T,B) and go to Procedure (II).
```

In what follows we are going to give a detailed description to the method for solving the sub-problem $SP(n)$.

1. Initialization. Construct the set S of all subsets of $\{1,2,\cdots,p\}$. Let $\mathcal{A} := \emptyset, \mathcal{B} := \emptyset, \mathcal{C} := \emptyset$ and $\mathcal{D} := \emptyset$.

2. Procedure (I). While $S \neq \emptyset$ do this step. For a subset $T \in S$, say, $T = \{1,\cdots,r\}$, analyze the real roots of the semi-algebraic system

$$\begin{cases} g_1 = y_0, \cdots, g_r = y_0, \\ g_{r+1} > y_0, \cdots, g_p > y_0, \\ y_0 < 1/9, 0 < y_1 < 1, \cdots, 0 < y_q < 1 \end{cases}$$

by using Gröbner Basis and Sturm's Theorem (this corresponds to the symbolic computation S1 in the Sketch Diagram). We compute the Gröbner Basis of $\{g_1 - y_0, \cdots, g_r - y_0\}$ with respect to an appropriate order of $\{y_0, y_1, \cdots, y_q\}$ to get a set of polynomials $G_0 \in Q[y_0, y_1, \cdots, y_q]$ such that

$$\text{RealZero}(\{g_1, \cdots, g_r\}|V) = \text{RealZero}(G_0|V)$$
$$(V = \{\frac{1}{12} < y_0 < \frac{1}{9}, 0 < y_1 < 1, \cdots, 0 < y_q < 1\}),$$

and G_0 belongs to one of the following three cases as shown in the Sketch Diagram.:

(1) $1 \in G_0$;
(2) A univariate polynomial of y_0 is contained in G_0 and G_0 is an ascending chain of polynomials, that is,

$$G_0 = \begin{cases} h_1(y_0), \\ h_2(y_0, z_1, \cdots, z_{l_1}), \\ h_3(y_0, z_1, \cdots, z_{l_1}, z_{l_1+1}, \cdots, z_{l_2}), \\ \cdots, \\ h_m(y_0, z_1, \cdots, z_{l_1}, z_{l_1+1}, \cdots, z_{l_2}, \cdots, z_q) \end{cases},$$

where $z_1, \cdots, z_q$ is a permutation of $y_1, \cdots, y_q$ and $1 \leq l_1 < l_2 < \cdots < q$.
(3) No univariate polynomial of y_0 is contained in G_0.

Here $\text{RealZero}(G|V)$ stands for the set of all real points $Y \in R^q$ such that $g(Y) = 0$ for all $g \in G$ and $v(Y) \neq 0$ for all $v \in V$. In S1, $\text{RealZero}(G|V)$ is obtained by reducing the Gröbner Basis G_b of the $\{g_1, \cdots, g_r\}$ using Maple function sturm(h,y,a,b), which returns the number of real roots in the interval (a,b] of polynomial h in y, as in the following procedures. The procedure reducing1 removes all univariate factors $h(y_0)$ of a polynomial g that has no real root in $(1/12, 1/9)$ and all univariate factors $h(y_i)(i = 1, \cdots, q)$ that has no real root in $(0, 1)$, reducing2 applies reducing1 to all members of a set of polynomials.

```
reducing2 := proc(G) map(reducing1, G) end proc
reducing1 := proc(g)
    prd(map(rdc, map2(op, 1, op(2, factors(g))))))
end proc
rdc := proc(g) local x, a, b;
    if nops(indets(g)) = 1 then
        x := op(1, indets(g));
        if x = y_0 then a := 1/12; b := 1/9
        else a := 0; b := 1
        end if;
        if sturm(g, x, a, b) = 0 or g = x - b then 1
        else g
        end if
    else g
    end if
end proc
prd := proc(a)
    if nops(a) = 0 then 1
    else op(1, a)*prd([op(2 .. nops(a), a)])
    end if
end proc
```

If $1 \in G_0$, then move this set T and all set $T' \in \mathcal{S}$ satisfying $T \subset T'$ to $\mathcal{A}$ (this operation is corresponding to ①).
If G_0 belongs to Case (3), then move T and all set $T' \in \mathcal{S}$ satisfying $T' \subset T$ into $\mathcal{B}$ (this is corresponding to ⑤).
If G_0 belongs to Case (2), then check if it is a strictly ascending chain, that is, it satisfies $m = q + 1$ and

$$
\begin{aligned}
&h_2 \in Q[y_0, z_1] \setminus Q[y_0], \\
&h_3 \in Q[y_0, z_1, z_2] \setminus Q[y_0, z_1], \\
&\cdots, \\
&h_m \in Q[y_0, z_1, \cdots, z_q] \setminus Q[y_0, z_1, \cdots, z_{q-1}],
\end{aligned}
$$

for some permutation $z_1, \cdots, z_q$ of $y_1, \cdots, y_q$. This is corresponding to S2 in the Sketch Diagram. If the answer is negative, then move T to $\mathcal{C}$ (which is corresponding to the Case (22) and operation ④). Otherwise, i.e., for Case (21), we compute the interval solutions $Ir(G_0)$ of G_0 using a Maple procedure IntervalRealRoot (explained later) and verify if there is any possible optimal solution with optimal tight indices T (this step is corresponding to S3). Since

$$
\mathrm{RealZero}(g_1 - y_0 = 0, \cdots, g_r - y_0 = 0 | V) = \mathrm{RealZero}(G_0 | V)
$$

and G_0 has finitely many solutions, we may use Gröbner Basis or Resultant computation together with reducing1 to create a univariate polynomial $\bar{h}_i(y_i)$ for each $i = 1, \cdots, q$ such that

$$
y \in \mathrm{Projection}_{y_i}(\mathrm{RealZero}(G_0 | V)) \implies \bar{h}_i(y_i) = 0.
$$

Let $\bar{h}_0 = h_0 \in G_0$. Then we modify `realroot` in Maple slightly to find all intervals solutions of a univariate polynomial h in a given interval $[a, b]$, for example, as follows.

```
intervalrealroot := proc(h, x, a, b) local a1, b1, r;
    a1 := sproot(subs(x = x - a, h), x);
    b1 := sproot(subs(x = b - x, h), x);
    r := min(gapofroot(h, x), min(a1, b1));
    a1 := realroot(h, min(1/2*r, 1/1024));
    r := [];
    for b1 in a1 do
        if a < op(1, b1) and op(2, b1) < b then
            r := [b1, op(r)]
        end if
    end do;
    r
end proc
sproot := proc(f, x) local r;
    r := 1;
    while 0 < sturm(f, x, 0, r) do r := 1/2*r end do;
    r
end proc
gapofroot := proc(h, x)
    sproot(prd(map2(op, 1,
    op(2, sqrfree(resultant(h, subs(x = x + u, h), x)))))/u,
    u)
end proc
```

Note that Maple command `map2(f, arg1, expr)` applies a function `f` to the operands or elements of `expr`, e.g.,

$$\text{map2}(\text{op}, 1, [[x1, y2], [x2, y2], [x3, y3]]) = [x1, x2, x3],$$

and `sqrfree` is the function for computing the square-free factorization of multivariate polynomials. Let

$$M_0 := \text{intervalrealroot}(\bar{h}_0, y_0, 1/12, 1/9),$$
$$M_1 := \text{intervalrealroot}(\bar{h}_1, y_1, 0, 1),$$
$$\cdots,$$
$$M_q := \text{intervalrealroot}(\bar{h}_q, y_q, 0, 1),$$

and

$$M_r := M_0 \times M_1 \times \cdots \times M_q.$$

Let $Ir(G_0) = \emptyset$. For each $I = (I_0, I_1, \cdots, I_q) \in M_r$, do the following interval computation (no division included) for all constraints

$$J_1 := g_1(I_1, \cdots, I_q) - I_0, \cdots, J_r := g_r(I_1, \cdots, I_q) - I_q.$$

If $0 \in J_i$ for all $i = 1, \cdots, r$, then push this I into $Ir(G_0)$. Let

$$Ir := \{I | I = (I_0, I_1, \cdots, I_q) \in Ir(G_0),$$
$$0 \in g_i(I) - I \text{ or } g_i(I) - I \subset (0, +\infty) \text{ for all } r + 1 \leq \cdots \leq p\}.$$

This finishes the computation S3. If $Ir = \emptyset$, then move T and all sets $T' \in \mathcal{S}$ satisfying $T \subset T'$ into $\mathcal{A}$ (corresponding to Case (211) and operation ②); otherwise, push (T, Ir) into $\mathcal{D}$ and remove T from $\mathcal{S}$ (which is corresponding to Case (212) and operation ③).

3. Procedure (II). While $\mathcal{B} \neq \emptyset$ do this step. Recall that each $T \in \mathcal{B}$, say, $T = \{1, \cdots, r\}$, satisfies that no Gröbner Basis of $G = \{g_1 - y_0, \cdots, g_r - y_0\}$ contains a univariate polynomial of y_0. As a preprocess of S4 we compute all Gröbner Basis of G with respect to all possible orders of $y_0, y_1, \cdots, y_q$. If any univariate polynomial $h(y_i)$ for some $j(1 \leq j \leq q)$ is generated in the process, do reducing1 defined in Step 2. If reducing1$(h) = 1$ then interrupt the computation for this T, move T and all $T' \in \mathcal{B} \cup \mathcal{C}$ satisfying $T \subset T'$ to $\mathcal{A}$. If reducing1$(h) \neq 1$ or no univariate polynomial is found, do LagrangeMultiplier for the following non-linear programming problem:

$$NLP(n, T):$$
$$\max y_0,$$
$$\text{s.t. } g_1(y_1, \cdots, y_q) - y_0 = 0, \cdots, g_r(y_1, \cdots, y_q) - y_0 = 0;$$
$$g_{r+1}(y_1, \cdots, y_q) - y_0 > 0, \cdots, g_p(y_1, \cdots, y_q) - y_0 > 0;$$
$$1/12 < y_0 < 1/9, 0 < y_1 < 1, \cdots, 0 < y_q < 1.$$

Here we show the outline of LagrangeMultiplier. Let

$$L = y_0 + \sum_{i=1}^{r} \lambda_i \cdot (g_i(y_1, \cdots, y_q) - y_0)$$

and

$$\frac{\partial L}{\partial y_0} = 1 - \lambda_1 - \cdots - \lambda_r,$$
$$\frac{\partial L}{\partial y_1} = \lambda_1 \cdot \frac{\partial g_1}{\partial y_1} + \cdots + \lambda_r \cdot \frac{\partial g_r}{\partial y_1},$$
$$\cdots,$$
$$\frac{\partial L}{\partial y_q} = \lambda_1 \cdot \frac{\partial g_1}{\partial y_q} + \cdots + \lambda_r \cdot \frac{\partial g_r}{\partial y_q}.$$

Then compute the Gröbner Basis of

$$\{g_1 - y_0, \cdots, g_r - y_0, 1 - \lambda_1 - \cdots - \lambda_q, \frac{\partial L}{\partial y_1}, \cdots, \frac{\partial L}{\partial y_q}\}$$

and remove factors that has no real root in the interval $(1/12, 1/9)$ for y_0 and $(0, 1)$ for $y_i(i = 1, \cdots, q)$ using reducing2. This finishes S4.

Let G_0 be the obtained result. Then, G_0 belongs to one of the following three cases.

(31) $1 \in G_0$;

(32) G_0 is a strict ascending chain of polynomials, say, $G_0 = \{h_0, h_1, \cdots, h_q\}$ such that

$$h_0 := h_0(y_0, z_1),$$
$$h_1 := h_1(y_0, z_1, z_2),$$
$$\cdots,$$
$$h_q := h_q(y_0, z_1, \cdots, z_q),$$

where $z_1, \cdots, z_q$ is a permutation of $y_1, \cdots, y_q$.

(33) $1 \notin G_0$ and G_0 is not a strict ascending chain for any permutation $z_0, z_1, \cdots, z_q$ of $y_0, y_1, \cdots, y_q$ with $z_0 = y_0$.

For Case (31) we move T and all $T' \in \mathcal{B} \cup \mathcal{C}$ which satisfies $T \subset T'$ to the set $\mathcal{A}$ (⑥), and for Case (33) move T to set $\mathcal{C}$ (⑨). If G_0 belongs to Case (32), then (starting S5) use Gröbner Basis or Resultant together with `reducing1` to get a univariate $\bar{h}_i$ for each $j = 1, \cdots, q$ that satisfies

$$\text{Projection}_{y_i}(\text{RealZero}(G_0|V)) \subset \{y|\bar{h}_i = 0\}$$

for each $j = 1, \cdots, q$, and use `intervalrealroot` to get the intervals solutions of $h_0, \bar{h}_1, \cdots, \bar{h}_q$:

$$M_0 := \texttt{intervalrealroot}(h_0, y_0, 1/12, 1/9),$$
$$M_1 := \texttt{intervalrealroot}(\bar{h}_1, y_1, 0, 1),$$
$$\cdots$$
$$M_q := \texttt{intervalrealroot}(\bar{h}_q, y_q, 0, 1).$$

Let $M_r := M_0 \times M_1 \times M_1 \times \cdots \times M_q$ and

$$Ir := \{I|I = (I_0, I_1, \cdots, I_q) \in Mr,$$
$$0 \in g_i(I) - I = g_i(I_1, \cdots, I_q) - I_0 \text{ for all } 1 \leq i \leq r, \text{ and}$$
$$0 \in g_i(I) - I \text{ or } g_i(I) - I \subset (0, +\infty) \text{ for all } r+1 \leq \cdots \leq p\}.$$

(S5 finished here). If $Ir = \emptyset$ then move T and all $T' \in \mathcal{B} \cup \mathcal{C}$ with $T \subset T'$ to $\mathcal{A}$ (this is corresponding to Case (321) and operation ⑦); otherwise, i.e., for Case (322), push (T, Ir) into $\mathcal{D}$ and remove T from $\mathcal{B}$ (⑧).

4. Procedure (III). While $\mathcal{C} \neq \emptyset$ do this step. Recall that for each $T \in \mathcal{C}$, say, $T = \{1, 2, \cdots, r\}$, the Gröbner Basis G_0 of

$$G = \{g_1(y_1, \cdots, y_q) - y_0, \cdots, g_r(y_1, \cdots, y_q) - y_0\}$$

after `reducing2` with respect to

$$V = \{\frac{1}{12} < y_0 < \frac{1}{9}, 0 < y_1 < 1, \cdots, 0 < y_q < 1\}$$

belongs to one of the following two cases.

(a) G_0 contains a univariate polynomial $h_0(y_0)$, and for some permutation $z_1, \cdots, z_q$ of $y_1, \cdots, y_q$,

$$G_0 = \left\{ \begin{array}{l} h_1(y_0), \\ h_2(y_0, z_1, \cdots, z_{l_1}), \\ h_3(y_0, z_1, \cdots, z_{l_1}, z_{l_1+1}, \cdots, z_{l_2}), \\ \cdots, \\ h_m(y_0, z_1, \cdots, z_{l_1}, z_{l_1+1}, \cdots, z_{l_2}, \cdots, z_q) \end{array} \right\};$$

(b) G_0 contains no univariate polynomial in y_0, that is, for some permutation $z_1, \cdots, z_q$ of $y_1, \cdots, y_q$, G_0 can be written as

$$G_0 = \left\{ \begin{array}{l} h_1(y_0, z_1, \cdots, z_{l_0}), \\ h_2(y_0, z_1, \cdots, z_{l_0}, z_{l_0+1}, \cdots, z_{l_1}), \\ \cdots, \\ h_m(y_0, z_1, \cdots, z_{l_0}, z_{l_0+1}, \cdots, z_{l_1}, \cdots, z_q) \end{array} \right\}.$$

In both cases, $m < q + 1$. The non-linear programming restricted on T becomes to the following form

$NLP(n, T)$:

max y_0,

s.t. $h_1 = 0, \cdots, h_m = 0$,

$g_{r+1}(z_1, \cdots, z_q) - y_0 > 0, \cdots, g_p(z_1, \cdots, z_q) - y_0 > 0$,

$1/12 < y_0 < 1/9, 0 < z_1 < 1, \cdots, 0 < z_q < 1$.

The problem of this form is a typical cylindrical algebraic decomposition problem and there are completed algorithms (like QEPCAD) for solving it. Since in the non-linear programming problems generated by Heilbronn optimal configurations, the constraints $f_1, f_2, \cdots, f_{35}$ are quadratic polynomials of $x_1, x_2, \cdots, x_{14}$, and all monomials of degree 2 are of $x_{2i-1} \cdot x_{2j}$ $(i = 1, \cdots, 7, j = 1, \cdots, 7; i \neq j)$ form, and hence can be considered as linear combinations of either $x_2, x_4, \cdots, x_{14}$ or $x_1, x_3, \cdots, x_{13}$. It is natural to expect that linear equations also existed in $\{h_1, \cdots, h_m\}$ and can be used to eliminate part of variables before applying CAD-based methods, as described in the following steps (S6).

(i) Find a subset, say, $\{z_1, \cdots, z_k\}$, of $\{z_1, \cdots, z_q\}$ such that $h_0, h_1, \cdots, h_m$ form a system of linear equations with variables $z_1, \cdots, z_k$. Let

$$z_1 = Q_1(z_{k+1}, \cdots, z_q), \cdots, z_k = Q_k(z_{k+1}, \cdots, z_q)$$

be the solution of the linear equations. It is clear that $Q_1, \cdots, Q_2$ are fractional polynomials.

(ii) Substitute $z_1 = Q_1, \cdots, z_k = Q_k$ into the constraints of $NLP(n, T)$, we get a new optimal problem with less variables.

max y_0,

s.t. $V_1(z_{k+1}, \cdots, z_q) = 0, \cdots, V_{m-k}(z_{k+1}, \cdots, z_q) = 0$,

$U_1(z_{k+1}, \cdots, z_q) > 0, \cdots, U_s(z_{k+1}, \cdots, z_q) > 0$,

$1/12 < y_0 < 1/9, 0 < z_{k+1} < 1, \cdots, 0 < z_q < 1$.

(iii) Reduce polynomials $V_1, \cdots, V_{m-k}, U_1, \cdots, U_s$ under the condition $1/12 < y_0 < 1/9, 0 < z_{k+1} < 1, \cdots, 0 < z_q < 1$. Namely, if $W|V_i$ or $W|U_i$, and $W \in Q[y_0]$ such that $\mathtt{sturm}(W, y_0, 1/12, 1/9) = 0$, or $W \in Q[z_j](k+1 \leq k \leq q)$ such that $\mathtt{sturm}(W, z_j, 0, 1) = 0$, then W (with possibly a change of sign) can be removed from the corresponding V_i or U_i.

If $1 \in \mathtt{Gr\ddot{o}bnerBasis}(V_1, \cdots, V_{m-k})$, or the set $L_1 > 0, \cdots, L_t > 0$ of all linear constraints in the reduced problem forms an empty set, or the solution y_0 of the following linear programming $LP(n, T)$

$$\max y_0,$$
$$\text{s.t. } L_1(y_0, z_{k+1}, \cdots, z_k) > 0,$$
$$\cdots,$$
$$L_t(y_0, z_{k+1}, \cdots, z_k) > 0,$$

satisfies $y_0 \leq 1/12$, then move T and all $T' \in C$ that satisfies $T \subset T'$ to A (this is corresponding to Case (41) and operation ⑩). Otherwise (i.e., for Case (42)), solve the following quantifier elimination problem $QE(n, T)$:

$$\exists y_0 \exists z_1 \cdots \exists z_q [V_1 = 0 \wedge \cdots \wedge V_{m-k} = 0 \wedge U_1 > 0 \wedge \cdots \wedge U_s > 0$$
$$(1/12 < y_0 < 1/9) \wedge (0 < z_{k+1} < 1) \wedge \cdots \wedge (0 < z_q < 1)],$$

where polynomials $U_i, V_j \in Q[y_0, z_{k+1}, \cdots, z_q]$ are reduced under $1/12 < y_0 < 1/9, 0 < z_{k+1} < 1, \cdots, 0 < z_q < 1$. This is S7 in the Sketch Diagram.

If the quantifier elimination $QE(n, T)$ has no solution (Case (421)), then again we move T and all $T' \in C$ that satisfies $T \subset T'$ to A (⑪), otherwise, push $(T, \mathtt{QEsolutions})$ into D and remove T from C (Case (422) and operation ⑫ in the Sketch Diagram), here QEsolutions is a set of the following $q + 1$ array of intervals $(I_0, I_1, \cdots, I_q)$ that satisfies $QE(n, T)$ and

$$I_1 = Q_1(I_{m+1}, \cdots, I_q), \cdots, I_k = Q_1(I_{m+1}, \cdots, I_q).$$

Computation shows that this preprocess is very useful since for the most $T \in C$, the simplified non-linear programming problem obtained in Step 4.3 has only one or two free variables, for which the quantifier elimination method is very effective. It is also found that for all $T \in C$, the quantifier elimination problems obtained in Step 4.3 have no solution. This means, all possible optimal configurations are contained in the set D generated in Steps 2 and 3.

5. Final Processing. Analyze the possible optimal configurations recorded in D. Computation shows that $A \neq \emptyset$ only for $n = 1$ and $n = 5$, which correspond to $(5C92)$ and $(5C81)$, respectively. In both cases, A has only one member. This means that there are at most two possible optimal configurations for Heilbronn's seven points in a unit square. It is easy to verified the following two properties.

(i) The configuration decided by (5C92) is symmetric to the configuration decided by (5C81) about the axis $y = x$;

(ii) Both configurations are congruent to the configuration suggested by Francesc Comellas, J. Luis A. Yebra in [1].

Therefore, we finally proved the conjecture on the optimal configuration for seven points in a square. The results we proved through computation reads as follows.

Theorem 2. *For any seven points contained in a unit square, the smallest area of the triangles formed by these seven points can not exceed $1/12$ if the convex hull of these seven points is a hexagon, and can not exceed the smallest positive root of*

$$u^3 + \frac{3}{38}u^2 - \frac{7}{76}u + \frac{1}{152} = 0,$$

which is, $u = 0.0838590090\cdots = 1/11.9247772\cdots$, if the convex hull is a pentagon.   Q.E.D.

5   Open Problems Related to the Heilbronn Configuration of Eight Points in the Square

Comellas and Yebra conjectured that the Heilbronn configuration of eight points in the unit square is the configuration as shown in the Fig. 10(a) and the smallest triangle is $(\sqrt{13} - 1)/36 = 0.072376\cdots$. The best configuration known before Comellas and Yebra was $(2 - \sqrt{3})/4 = 0.066987\cdots$ given by Goldberg in 1972, as shown in the Fig. 10(b).

Let $\{p_1, p_2, \cdots, p_8\}$ be any eight points in the unit square. It is easy to prove that if the convex hull of these points is a triangle or a tetragon, then the smallest area of triangles formed by these points is not great than $h_{4,4} = (2 - \sqrt{3})/4$ in

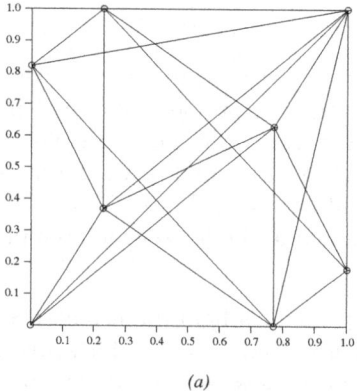

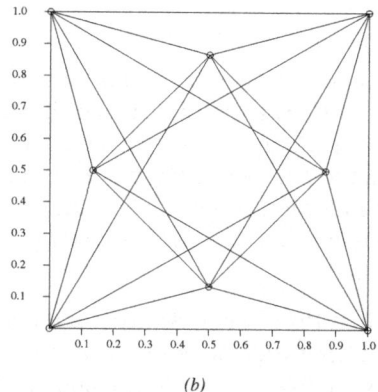

(a)   (b)

Fig. 10. Comellas and Yebra's conjecture and Goldberg's on the Heilbronn Configuration of eight points in the square

view of Lemma 1. If the convex hull is a pentagon, say, $p_1 p_2 \cdots p_5$, and therefore, $p_6, p_7, p_8 \in p_1 p_2 \cdots p_5$, then it is also easy to prove that either one of $p_i p_j p_k (0 \leq i < j < k \leq 5)$ contains two points of p_6, p_7, p_8, or each of the following five triangles

$$p_1 p_2 p_3, p_2 p_3 p_4, p_3 p_4 p_5, p_4 p_5 p_1, p_5 p_1 p_2$$

contains one of p_6, p_7, p_8. Thus, we have also

$$h_{5,3} = \min\{\text{area}(p_i p_j p_k) | 0 \leq i < j < k \leq 8\} \leq \max\{\frac{2 - \sqrt{3}}{4}, \frac{1}{3} \cdot \frac{\sqrt{3}}{9}\} = \frac{2 - \sqrt{3}}{4}.$$

On the other side, it is known (see [10]) that for any convex octagon $p_1 p_2 \cdots p_8$ the following inequality holds:

$$\min\{\text{area}(p_i p_j p_k) | 0 \leq i < j < k \leq 8\} \leq \frac{2 - \sqrt{2}}{8} \cdot \text{area}(p_1 p_2 \cdots p_8),$$

and, in any convex octagon $p_1 p_2 \cdots p_8$ there exists a convex octagon (called H-octagon) $q_1 q_2 \cdots q_8$ such that

$$\text{area}(q_{l_1} q_l q_{l+1}) = \min\{\text{area}(p_i p_j p_k) | 0 \leq i < j < k \leq 8\}, l = 1, 2, \cdots, 8,$$

here $q_0 = q_9, q_9 = q_1$. Since $(2 - \sqrt{2})/8 = 0.073223 \cdots$ is slight larger than $(\sqrt{(13)} - 1)/36 = 0.072376 \cdots$, we can not simply prove that all convex octagons $p_1 p_2 \cdots p_8$ contained in the unit square satisfy

$$\min\{\text{area}(p_i p_j p_k) | 0 \leq i < j < k \leq 8\} \leq \frac{\sqrt{13} - 1}{36}.$$

For this, we have the following open problem.

Open Problem 1. *Prove that if $p_1 p_2 \cdots p_8$ is a convex octagon contained in the unit square, then*

$$h_{8,0} = \min\{\text{area}(p_i p_j p_k) | 0 \leq i < j < k \leq 8\} \leq \frac{1}{16}.$$

This problem can be reduced to prove that $q_1 q_2 \cdots q_8$ is an H-octagon, that is,

$$\text{area}(q_{l-1} q_l q_{l+1}) = a,$$

and $ABCD$ is the smallest parallelogram containing $q_1 q_2 \cdots q_8$, then $a \leq 1/16 \cdot \text{area}(ABCD)$. Using similar analysis to Lemma 2 we can prove that if $ABCD$ is the smallest parallelogram containing an H-octagon, then the collocation of $q_1 q_2 \cdots q_8$ with $ABCD$ must be the case as shown in the following Fig.11 up to a permutation of indices. Note that the following relations

$$q_1 q_2 // q_8 q_3, q_2 q_3 // q_1 q_4, \cdots, q_8 q_7 // q_6 q_1.$$

hold for all H-octagons $q_1 q_2 \cdots q_8$.

For a complete proof to the conjecture of Comellas and Yebra on Heilbronn configuration of eight points in the unit quare, we have the following open problem.

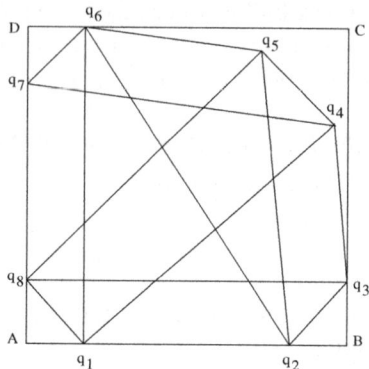

Fig. 11. The smallest parallelogram covering an H-octagon

Open Problem 2. *(a) Prove that if $p_1 p_2 \cdots p_6$ is a convex hexagon contained in the unit square, and p_7, p_8 are contained in the hexagon, then*

$$h_{6,2} = \min\{\text{area}(p_i p_j p_k)|0 \le i < j < k \le 8\} \le \frac{\sqrt{13}-1}{36}.$$

(b) Assume that $p_1 p_2 \cdots p_7$ is a heptagon contained in the unit square, and p_8 a point contained in the heptagon. Compute the maximal minimum

$$h_{7,1} = \max\min\{\text{area}(p_i p_j p_k)|0 \le i < j < k \le 8\}.$$

The numerical searching for the optimization configuration in the case (b) of the Open Problem 2 shows that $h_{7,1} \ge 0.067108\cdots$, the positive real root of the following cubic equation:

$$45796x^3 + 12714x^2 - 225x - 56 = 0,$$

as shown in Fig.12, which is slightly larger than $h_{4,4} = (4-\sqrt{3})/4 = 0.066987\cdots$, the optimal value in Fig.10(b).

The following open problem is an analogue to Lemma 7 for finding the optimal configuration of eight points whose convex hull is of unit area such that the smallest area of triangles formed by the eight points is maximal.

Open Problem 3. *Let $p_1, p_2, \cdots, p_8$ be any eight points in the plane such that the area of the convex hull of $p_1, p_2, \cdots, p_8$ is 1. Find the maximal value H_8 of the smallest area of triangles formed by $p_1, p_2, \cdots, p_8$.*

Our conjecture to the answer of the last problem is that

$$H_8 = \frac{1}{14\cos(\pi/14)} = 0.079279\cdots,$$

and the optimal configuration is formed by the seven vertices of an affine regular heptagon and the center of the heptagon, as shown in the Fig. 13.

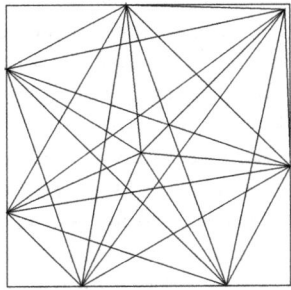

Fig. 12. A local optimization configuration for $h_{7,1}$

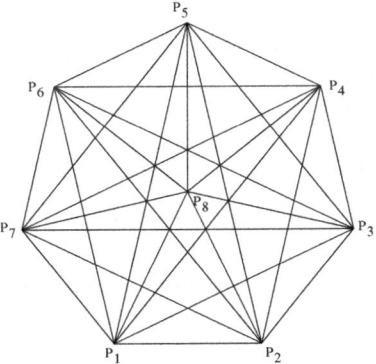

Fig. 13. An affine convex octagon with its center

6   Conclusion

We presented in this paper a proof to the Conjecture of Comella and Yebras on the Heilbronn configuration of seven points in a unit square. We first investigated the combinatorial types of the possible optimal configurations and the collocation of (the convex hull of) the seven points with the edges of the square, and therefore reduced the proof to solving of 226 non-linear programming problems of the following form:

$$\max \; x_0,$$
$$\text{s. t. } \; S_{i,j,k}(x_1, \cdots, x_q) \geq x_0, (1 \leq i < j < k \leq 7)$$
$$0 \leq x_1, \cdots, x_q \leq 1$$

where $x_1, \cdots, x_q$ are the coordinates of the seven points, $S_{i,j,k}$ are quadratic polynomials of $x_1, \cdots, x_q$ and satisfy degree($S_{i,j,k}, x_l$) ≤ 1 for all $x_1, \cdots, x_q$. The analysis significantly reduced the number of unknown coordinates from 14 to $6 \leq q \leq 8$. For the enumeration of combinatorial types formed by seven points in a square the readers may get general information from [1,2]. Then we worked to reduce the number of quadratic constraints in the 226 non-linear programming problems. In the third stage we used symbolic computation to get strict proof to that configuration suggested by Comellas and Yebras is the unique solution (up to congruent) to the non-linear programming problems by regarding them as quantitative eliminations. We also analyzed the possible ways to prove the Conjecture of Comellas and Yebra on eight points. Our impression is that the symbolic computation may have potential for this job if there are more effective methods to search the combinatorial types of the possible optimal configurations, to reduce the number of unknowns, and to simplify the constraints of non-linear programming problems associated with the optimal configurations.

References

1. Aichholzer, O., Aurenhammer, F., Krasser, H.: Enumerating order types for small point sets with applications. Order 19(3), 265–281 (2002)
2. Aichholzer, O., Krasser, H.: Abstract order type extension and new results on the rectilinear crossing number. Computational Geometry: Theory and Applications, Special Issue on the 21st European Workshop on Computational Geometry 36(1), 2–15 (2006)
3. Comellas, F., Yebra, J.L.A.: New lower bounds for heilbronn numbers. Electr. J. Comb. 9(6), 1–10 (2002)
4. Dress, A.W.M., Yang, L., Zeng, Z.: Heilbronn problem for six points in a planar convex body. In: Combinatorics and Graph Theory 1995, vol. 1 (Hefei), pp. 97–118. World Sci. Publishing, Singapore (1995)
5. Goldberg, M.: Maximizing the smallest triangle made by n points in a square. Math. Magazine 45(3), 135–144 (1972)
6. Yang, L., Zeng, Z.: Heilbronn problem for seven points in a planar convex body. In: Dingzhu, D., Pardalos, P.M. (eds.) Minimax and Applications (1995)
7. Yang, L., Zhang, J., Zeng, Z.: On exact values of heilbronn numbers for triangular regions. Tech. Rep. 91-098, Universität Bielefeld (1991)
8. Yang, L., Zhang, J., Zeng, Z.: On goldberg's conjecture: Computing the first several heilbronn numbers. Tech. Rep. 91-074, Universität Bielefeld (1991)
9. Yang, L., Zhang, J., Zeng, Z.: A conjecture on the first several heilbronn numbers and a computation. Chinese Ann. Math. Ser. A, 13, 503–515 (1992)
10. Zeng, Z., Shan, M.: Semi-mechanization method for an unsolved optimization problem in combinatorial geometry. In: Proceedings of the 2007 ACM Symposium on Applied Computing, pp. 762–766. ACM, New York (2007)

Author Index